The Humex Book of PROPAGATION

John Harris

Macdonald

Contents

© Macdonald Educational Ltd, 1980
Published in association with
Humex Ltd
First published 1980
Macdonald Educational Ltd
Holywell House
Worship Street
London EC2 2EN

ISBN 0 356 07156 1 (paperback edition)
ISBN 0 356 07157 X (cased edition)

Printed and bound by
Purnell & Sons Ltd
Paulton

Editorial manager
Chester Fisher
Editors
Brenda Clarke
Neil Tennant
Designer
Peter Benoist
Picture Research
John Harris
Production
John Moulder

Introduction

Raising your own plants is far more satisfying than having to select from the mediocre stock of many trade nurserymen. Besides which, you are likely to save a small fortune. Even now, for an outlay of a few pence on a packet of seed, it is possible to produce plants that would cost pounds to buy. In addition to propagation by seed, most plants can be increased simply and quickly by a host of vegetative methods. In terms of skill required, plant propagation, unlike most aspects of gardening, is not surrounded by muck and mystery: it is very much an exact science. To be successful, therefore, it will be necessary to learn a little about the way plants function, master a few basic skills and, if you have not already done so, invest in some equipment. Once these are acquired, you have opened the way to one of the most creative and profitable of pastimes— the perpetuation of plants.

Aids and equipment

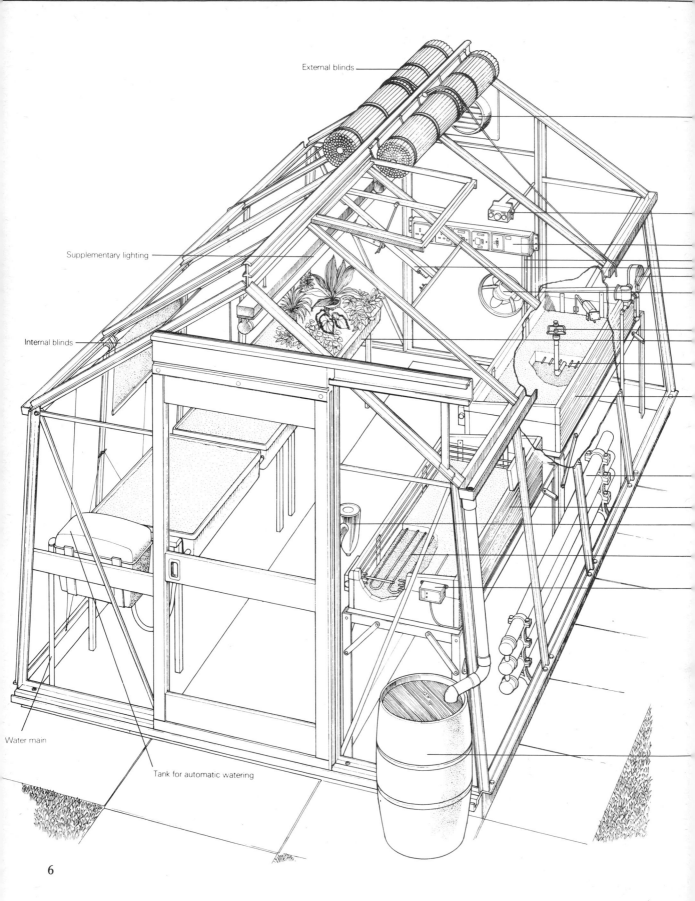

External blinds

Supplementary lighting

Internal blinds

Water main

Tank for automatic watering

Extractor fan

Heater thermostat

Control panel
Electricity cable
Automatic vent opener
Pump

Air circulator

Mist nozzle
Capillary bench

Soil warming cable

Mist propagator

Electric tubular heater

Heated bench propagator

Electric fumigator

Soil-warming cable

Thermostat

Water storage

Having decided to raise plants at home, a gardener will need some basic equipment with which to start. And before investing money in propagating equipment he needs to decide where to install it. Something small and simple for raising a couple of dozen plants can be stood on the kitchen window-sill, but anything more sophisticated will need a greenhouse. So this may very well be the first piece of propagating equipment to buy.

Greenhouses

Regardless of make, the best type of greenhouse for propagation is one made of glass—this is a far better heat-insulating material compared with most plastics used for glazing. For practical purposes there is not much to choose between a timber or aluminium frame. Fittings are more easily attached to the former, while the latter has an almost indefinite life. Again, cladding the lower half of the walls will reduce winter fuel bills, but will also restrict growing space in summer. It becomes a case of deciding which is best for you.

However, it is essential to ensure that the greenhouse is firmly anchored to the ground on a solid foundation, and that the glazing is reasonably air-tight, with no gaps through which heat can be lost or the wind can blow. Ventilators, however, should be large enough to keep the temperature in the greenhouse effectively under control when the need arises. The vent area should equal at least one fifth that of the floor. Small greenhouses even

Left: The ideal greenhouse in which to propagate plants has a mist propagating bench; heated bench; automatic watering; fumigator; lighting; automatic vent opener and extractor fan; and blinds to give summer shade and winter frost protection. A fan keeps air moving to prevent still, cold conditions during winter which can give rise to disease. The greenhouse floor is solid—made of concrete or paving slabs.

Above: A circulating fan is useful to keep the air within the greenhouse constantly moving. It therefore helps to prevent damping-off.

Below: A thermometer giving both maximum and minimum temperatures is an essential piece of equipment in a greenhouse.

then will be prone to overheating and as a safeguard an extractor fan, thermostatically controlled, can be installed in the end wall.

Electric heating is preferable to other methods as it offers accurate temperature control. Gas is a close second. Solid fuel and paraffin heaters are both difficult to regulate and need constant refuelling.

Mains water laid directly into, or just outside, the greenhouse will make the installation of automatic watering and sophisticated propagating equipment a much simpler task, eliminating the need for bulky water storage tanks. However, keep a butt outside to collect rainwater as this is invaluable for watering lime-hating plants if you live in a hard water area.

Blinds serve a double function and this justifies the high cost. Besides keeping out the sun in summer, they provide valuable frost protection on cold winter nights, adding a second skin to glazing. Plastic sheeting lining the inside of the greenhouse in winter may conserve heat, but it also reduces light and causes high levels of condensation, an important contributing factor to disease problems.

Needless to say, a permanent, properly installed electric power supply is almost essential. Pages 16–17 explain how this should be put in. Do *not* string a cable from the nearest power point in the home down the garden—this is both illegal and highly dangerous.

Besides staging, other essential pieces of equipment are: a maximum and minimum thermometer, to tell you not only what the temperature goes up to in the day but how cold it gets at night; and a small fan to circulate air. On cool days, when vents are closed, this will keep the air moving, preventing close, humid conditions under which many diseases thrive.

Propagators

Owning a self-contained propagator is like having a hot-house in miniature. Much more important, it also provides the right conditions for raising a wide range of plants from seeds or cuttings, even in an unheated greenhouse. Heat is applied efficiently, directly to the material being propagated. It is therefore possible to keep the overall temperature of the greenhouse low, or to shut off the heat completely. For example, if a minimum greenhouse temperature of 7°C (45°F) is maintained—as it can be quite easily by using a thermostatically-controlled electric heater—an electric propagator can give the higher temperature of 18–21°C (65–70°F) needed to raise most plants. A first step is to look at the types of propagators for sale off the shelf.

Above: Plants raised with the aid of a simple paraffin-heated propagator. A small burner heats a flat water-filled tank on which stand seed trays and pots.

Plastic tops and polythene bags

The cheapest way to supply the high humidity needed for rooting cuttings and germinating seeds is to cover pots and trays with clear, rigid plastic domes, or with polythene bags supported on galvanized wire frames. Compost then reaches quite a cosy temperature, and with the high humidity it is possible to raise a wide variety of plants in a cool or unheated greenhouse from late spring onwards. Of course, there is little or no protection during cold spells or at night, and it is here that the added safeguard of a heated propagator will pay dividends.

Bottom heat . . .

If you intend to raise only a few plants, then a simple heated base may be more than adequate for your needs. This consists of an insulated base containing a mains heating element that will give a temperature lift of around 11°C (20°F). Therefore, if the greenhouse is at 7°C (45°F) the propagator will be able to maintain a temperature of 18°C (65°F). If buying a heated base for use in a cold greenhouse, bear in mind that the rise from a low night temperature to that required for propagation could be too great for the heating element to

Above: Enclosing cuttings within a polythene bag, supported on a wire frame, will provide the heat and high humidity needed for the rapid rooting of most softwood cuttings.

Below: A sophisticated electric propagator, fitted with a thermostat. This enables the soil temperature to be adjusted according to the needs of the plants being raised.

cope with. The heating element must be distributed evenly over the base, right up to the edges, to produce an overall gentle heat without hot spots. Bases are normally large enough to take a standard sized seed tray, and are ideal for the beginner who wants to experiment economically, or for the window-sill gardener. If your greenhouse has no electricity, a simple paraffin heated propagator is the answer. In this, a small burner heats a flat, water-filled tank upon which stand seed trays and pots.

... with temperature control

All these propagators are electric and fitted with thermostats to adjust their temperature according to that of the greenhouse and the needs of the plants being raised. A thermostat should have a wide temperature control range—at least 10–27°C (50–80°F)—to cope with everything from the propagation of alpines to that of cacti and orchids. Flat electrically-heated bases fitted with thermostats are useful, but for a small extra cost it is worth investing in a large heated tray which, if desired, can be filled with compost in which to strike cuttings or place trays and pots. The simplest covering for a tray-type propagator is a polythene tent supported by a galvanized wire frame: this will maintain a

Below: Seedlings and young plants gaining the warmth provided by a simple electrically-heated base. Large enough on which to stand one full-sized seed tray, the heated base is therefore ideal for the window-sill gardener, or for the beginner who wants to experiment.

warm, humid atmosphere. In a more expensive version, the base can be surmounted by an aluminium-framed structure, glazed with rigid plastic and fitted with sliding doors to allow easy access to plants.

The ultimate off-the-shelf propagators are fitted with sophisticated hoods and thermostatically-controlled air warming, but for those contemplating such an outlay, it is seriously worth considering investing in professional equipment such as mist propagators and heated benches.

The emphasis so far has been on maintaining temperature and humidity, but the dangers of overheating must not be ignored. On a sunny day, even in winter, the temperature within a closed propagating case will rise rapidly and may cause scorching of tender seedlings and cuttings. A simple precaution is to drape the propagator with a special horticultural shade netting or coloured translucent polythene.

An electric propagator also provides an inexpensive way of overwintering hothouse plants

such as orchids, which need a high temperature, although a careful watch must be kept for rot disease.

Heated benches

When a gardener wants to propagate plants on a large scale, it is far cheaper to assemble a propagating bench using soil-warming cable than to buy a ready-made unit. Soil-warming cables operate from the mains and must be wired to an insulated fused socket. The safest type of cable for an amateur to buy is one which has a wired-in thermostat and is waterproof. Cable kits are made to heat a specific area. For example, a 75-watt cable, measuring 6 m (20 ft) in length, will provide sufficient bottom heat for a bench area of 0.7 m^2 (7½ sq ft). Therefore 10 watts of heat—equal to 800 mm (2½ ft) of cable—will be required to heat 0.1 m^2 (1 sq ft) of bench. With a minimum greenhouse temperature of 7°C (45°F) such heat output will maintain a comfortable propagating temperature of 18°C (65°F). The thermostat should be of the rod type and have a temperature control range of at least 5°C (40°F) to 23°C (75°F). The thermostat is sited in the bench, so the rod (the temperature sensor) lies just below the surface of the sand.

Cable can be bought screened or unscreened. It is best to buy the screened variety as—when properly earthed—there is no risk of receiving a shock should you accidentally cut through it.

Pages 12–13 show how to construct a simple heated bench, but a few points concerning construction need to be emphasized. The base of the bench must have drainage holes to prevent waterlogging. When laying the heating cable make sure there are no kinks, sharp bends or overlapping. Paint the inside walls of the propagator white to intensify light and promote rapid, sturdy growth. A glass top will help to maintain a humid, warm atmosphere.

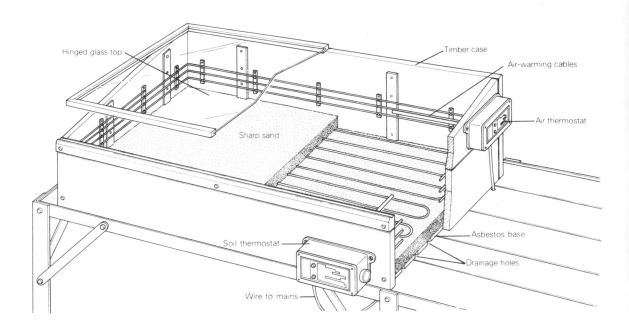

Above: Components of a heated bench fitted with soil- and air-warming cable.

Below: Components of a mist propagating unit. Note that the pressure supplied must be at least 30 psi in order to operate the solenoid valve. If pressure falls below this, water will have to be pumped to the system.

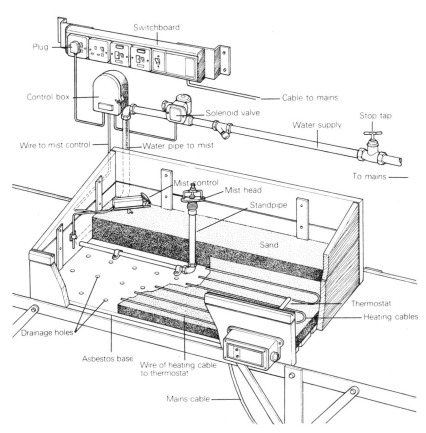

Air warming

If the heated bench has high walls, the installation of air-warming cable will make it possible to propagate more varieties, including tropical plants which like warm tops as well as warm bottoms. As with soil-warming cable, an air-warming cable should be thermostatically controlled. The cable is attached to the inside walls of the propagator by small stainless steel clips. Of course, there is no point in installing air warming unless the propagator is fitted with glass lights to keep the heat in, but ensure that these can be opened to allow for ventilation.

The table below shows the standard cable lengths that can be bought with thermostats ready-wired. It also gives their wattage and the areas they will heat without becoming overloaded while maintaining a constant temperature of around 21°C (70°F).

Cable watts	Cable length		Max. area heated soil		air	
	m	(ft)	m²	(ft²)	m²	(ft²)
75	6	(20)	0.7	(7½)	0.5	(5)
150	12	(40)	1.4	(15)	1.0	(10)
300	24	(80)	2.8	(30)	1.8	(20)
500	40	(133)	4.6	(50)	3.0	(33)

Mist propagation

When a leafy cutting is taken, it immediately starts to wilt and die simply because it has no roots to replenish food and, more important, water. In order to make a cutting form roots, water loss, which mainly occurs through the leaves, must be controlled. Placing cuttings in a closed propagator goes some way towards achieving this, but at the same time the hot, humid atmosphere it supports also provides the perfect environment for many diseases. In addition, shading, which is necessary in sunny weather, reduces photosynthesis (food production) and therefore a cutting's chance of rooting. Covering the leaves of cuttings with a thin film of water prevents wilting, and controls the surrounding temperature so that shading is unnecessary. The higher light intensity means that a cutting can continue making more food than it requires for its immediate needs. This in turn encourages rapid root initiation and development.

To be effective, misting must be used together with soil warming, and instead of the 10 watts per 0.1 m² (1 sq ft) required for a simple heated bench, 15 watts per 0.1 m² (1 sq ft) are needed to keep a mist bench adequately heated. As a mist unit is not enclosed, there is no advantage to be gained from installing air warming.

Each mist nozzle will cover an area 1 m (3¼ ft) in diameter with spray, so it is possible to root 200 or more cuttings at a time under a single nozzle. Water is supplied through a feed-pipe and the nozzle is held about 480 mm (18 in) above the bench on a standpipe. Water to the mist nozzle is switched on and off by means of a simple solenoid valve. This is activated in turn by a detector measuring the amount of water the cuttings need. A minimum water pressure of 30 psi is required to activate most solenoid valves. If a suitable mains supply is not available, water must be fed to the system through

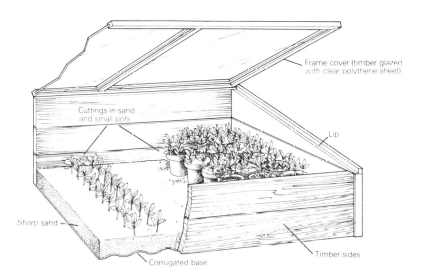

Above: Components of an unheated closed case. The light, fitted with glass, will allow humidity on the bench to be maintained at a high level if the cuttings are sprayed regularly. However, without soil-warming cables, unless the greenhouse is kept reasonably warm all the time, the night temperature within the frame will drop quite low between late autumn and early spring.

a pump. Before it reaches the solenoid valve, the water must be filtered, as even very small particles can easily block the nozzles. The bench must be level, all standpipes vertical, and the nozzles kept scrupulously clean.

As mentioned earlier, the frequency of misting is controlled automatically by a detector which measures the quantity of water cuttings need. There are three detectors to choose from.

Electro-mechanical balance detector works on a simple balance principle. A foam pad absorbs moisture while cuttings are being sprayed. Eventually the weight of the absorbed water tips the balance arm, causing a switch at the other end of the arm to break contact and cut off the spray. Conversely, as moisture evaporates from the cuttings it is also lost from the foam pad at a similar rate. Thus, the pad grows

lighter and eventually tips back again, switching on the mist. The degree of wetting can be controlled by moving the detector along its mounting arm, either increasing or reducing the amount of water required to tip it. The rate of mist can be further reduced to 'wean' cuttings by placing immediately below the absorbent pad a small heater which is operated by a switch in the control unit.

Electronic leaf detector consists of a piece of plastic incorporating two terminals. This is placed on the mist bench among the young cuttings. As moisture evaporates from the cuttings, the terminals dry correspondingly, so breaking an electric circuit which is a signal for the mist to switch on. Conversely, as the terminals grow damp the water is switched off.

Solar control detector. Here, light intensity in the greenhouse controls frequency of misting. A photo-electric cell absorbs light, and when a pre-determined level is reached it activates the mist. The sunnier the weather, the more frequently the mist is switched on. A simple over-ride control ensures that the mist sprays periodically when it is dark.

The rooting medium in a mist bed must be extremely well drained, and holes 13 mm (½ in) in diameter should be drilled at 150-mm (6-in) intervals in the base.

Apart from encouraging the rapid rooting of most cuttings, mist is used successfully by commercial nurserymen to germinate many sorts of seed. After sowing, boxes and pots are placed on the mist bench and removed as soon as seedlings start to appear. However, misting does have certain limitations. Plants with hairy leaves, begonias for instance, will rarely tolerate being continually wet and usually rot.

It is most important to ensure that the control unit and all electrical equipment are kept well away from the mist. Water is always present and therefore electrical installations must be fitted with the greatest of care by a competent, well-qualified electrician.

Constructing a heated bench

1
Constructing the frame of a heated bench. It should be made from 13-mm (½-in) waterproof plywood or a similar timber, and all fittings (e.g. clips, screws, nuts, bolts, etc.) should be galvanized or made of stainless steel to avoid rust. The timber ideally needs to be treated with a horticulturally safe preservative, prior to painting.

2
On a mist bench good drainage to prevent waterlogging of the rooting medium is absolutely essential. Holes 13 mm (½ in) in diameter should be drilled through the asbestos base at 150-mm (6-in) centres. If the bench is to be used with soil-warming cable only, the holes can be spaced at 220–300-mm (9–12-in) centres.

6
The heating cable is carefully uncoiled and passed through a small hole by the side of the thermostat control box. A rubber grommet is used to seal the lead tightly in the hole. The cable must be laid evenly over the surface, taking care to avoid leaving any sharp bends or kinks which may later result in overheating and damage.

7
If a mist nozzle is to be added, plastic pipe must be supplied to conduct the water. The piping must be firmly fixed to the base of the bench by saddle clips. Water flow is controlled by a solenoid valve. This is activated by a detector which automatically measures the rate of water lost from cuttings by evaporation.

8
The mist nozzle is fitted onto a standpipe which rises about 450 mm (18 in) above the base of the bench. Brass nozzles are preferable to plastic ones. Each nozzle will cover an area 1 m (3¼ ft) in diameter with spray, so making it possible to root 200 or more cuttings beneath every one.

3
Soil-warming cable has to be laid on a prepared bed of coarse, washed river sand. It is important to use this grade rather than the soft builders' sand, peat, ash, or any other kind of material that tends to waterlog readily or may contain noxious substances.

4
The bench must be fitted with a layer of sand 40 mm (1½ in) deep. This is spread over the base and then levelled off, ready to act as a bed on which the soil-warming cable will be laid.

5
The thermostat is fitted with the rod just below the surface of the sand, and is held in position by two screws. The rod (the temperature sensor) must be set parallel with the base, pointing neither upwards nor down. The thermostat needs a temperature range of at least 5–21°C (40–70°F).

9
An electro-mechanical detector to control the rate of misting is suspended over the cuttings. At the back, fitted to the greenhouse wall, are the control box, solenoid valve and water filter. At all times, the nozzle's pin-sized opening must be kept scrupulously clean.

10
Bench fitted with soil- and air-warming cables. It is covered with a glass light to maintain a warm, humid atmosphere.

Artificial lighting

Plants need light as a source of energy to convert carbon dioxide (taken in through the leaves), water and minerals (both absorbed by the roots) into sugars which provide a stable but easily-used food source. This, basically, is the process of photosynthesis. For most of the year natural daylight is sufficient, but during late winter and early spring—the start of the propagating season—it often drops below the desired level for optimum growth of cuttings, seedlings and young plants. It then becomes necessary to boost natural daylight with artificial light.

What type of light do plants require? Natural white light is composed of different colours—collectively called the spectrum—and green plants make use of only the blue and red parts. For plant nutrition the blue part is most important.

Do remember, however, that for artificial light to be of value, all the other factors controlling growth must be adequately supplied. For example, there is no point in installing an expensive lighting system if growth will be restricted by low temperatures.

Types of light

Household tungsten filament bulbs may be cheap but for propagation they are ineffective, as plants grown under them tend to become leggy. The most suitable and readily-available lamps for use by the amateur are fluorescent and mercury fluorescent varieties. These also have a very long life, which helps to justify the higher cost.

Fluorescent tubes. Plants respond best to the 'warm light' and 'Grolux' types. Unlike mercury fluorescent lamps, these do not produce much heat and can be suspended close to plants, even seedlings, without risk of scorching. All fittings must be waterproof, and tubes installed beneath reflectors so that maximum light is thrown down on to plants.

Mercury fluorescent. This type of lamp provides high-intensity light that is readily used by plants for photosynthesis. Complete fittings purpose-designed for greenhouse installation should be bought. These consist of bulb (normally 160 watt), reflector, lamp-holder and suspension hook.

Mercury fluorescent lamps of 160 watts are suspended about 750 mm (2½ ft) above benches at 750-mm intervals. Each lamp will then provide a very even light over an area of about 0.6 m^2 (6½ sq ft). Fluorescent tubes can be strung 450 mm (1½ ft) above benches. Fitted with reflectors, each will satisfactorily light a strip about 375 mm (15 in) wide along its entire length.

Switching on and off can be done by hand, or by a time switch similar to the type used on central heating systems.

Propagation under lights

Bedding plants
Sow seed in the normal manner and germinate in a heated closed propagator or under a mist unit. As soon as seedlings start to emerge, transfer them under lighting where optimum light level can be maintained for 16 hours a day. Seedlings develop much faster, are sturdier and more uniform than their 'unlit' counterparts. This means that they can be pricked out sooner, so reducing the risk of damping off.

Begonias from seed
Transfer seedlings under lights as soon as they start to emerge and apply the same treatment as for bedding plants.

Cacti and other succulents
Early spring sowings ensure well-established young plants by the following autumn.

Maintaining optimum light levels for 14 hours a day from germination to early summer prevents weak, soft growth.

Cucumbers
Early crops. Four to six extra hours of lighting applied to early sowings for the four weeks after germination will produce sturdy, short-jointed plants at least a week ahead of plants raised by conventional methods.

Cuttings
Soft, semi-ripe and hardwood. Three to four hours of extra lighting applied to all cuttings planted between autumn and late spring will encourage more rapid rooting. This greatly reduces losses from rot disease caused by unrooted cuttings standing in wet compost. Autumn-struck pelargonium ('geranium') cuttings are particularly responsive to treatment.

Lettuce
Winter- and early spring-cropping varieties. For the first couple of weeks after germination, light kept at the optimum level for 12 to 14 hours a day will advance harvesting by two weeks or more. However, longer treatment can result in bolting (plants running to seed).

Tomatoes
Early and late crops. Germinate seed in a propagator or under mist. Prick out seedlings, as soon as they are large enough to handle, into 100-mm (4-in) pots and place under lamps where optimum light levels are maintained for 16 hours a day. Plants are left under lamps until they need spacing out—after two to three weeks. This treatment brings on plants to the five-week stage in three weeks.

To encourage maximum plant growth, it is necessary to supplement natural light in early morning and late afternoon. A typical programme for early spring would be:

lights on 6am–9am
lights off 9am–3pm
lights on 3am–7pm

When the weather is very overcast lights can be left on all day. For any plant, the period of light at the optimum level need not exceed 16 hours in any one day.

The length of time plants are left under lighting will depend upon the space available. The chart below gives some ideas for treatments and their advantages.

Heating frames

A cold frame has only limited uses in propagation: hardening off plants before they go out in the open; the raising of some hardy plants; and providing frost protection to slightly tender subjects. To keep warm it relies on the heat of the sun. However, converted to a hotbed its usefulness will be greatly extended. With permanent frost protection gained by soil-warming, it can be used for almost all the same functions as a small heated greenhouse, but at a fraction of the cost.

The ideal type of frame for general propagation should be constructed of timber, with glass lights. Glass and metal-sided frames do not retain heat well and, generally, are too flimsy. A reasonable amount of headroom is needed—300 mm (12 in) at the front and 450 mm (18 in) at the back at least—and the frame light should be close-fitting. Although heating will be installed, the fuel bill can be reduced by siting the frame in a sunny, sheltered position. The ground beneath must be well-drained. Treat all timber with a non-toxic horticultural-grade preservative—*not* creosote!

The diagram shows a frame with soil-warming cable installed, but a few points are worth noting. The cable must be laid on a flat,

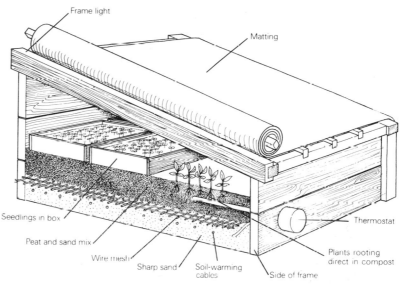

level surface of material such as perlite, vermiculite or gritty sand. A thin layer—about 25 mm (1 in)—of compost or sand is spread over the top, followed by a wire mesh screen to give the cable some protection. (Although you will not receive a shock by accidentally cutting through a properly earthed cable, it is expensive to replace.) The length of cable should be sufficient to give a heat output of around 7 watts per 0.1 m² (1 sq ft). Heating cables are sold in standard lengths which are completely sealed and insulated, and therefore must not be shortened. The table gives standard cable lengths and the largest sized frame each will heat.

Cable watts	Cable length		Max. area heated			
			soil		air	
	m	(ft)	m²	(ft²)	m²	(ft²)
75	6	(20)	1.0	(11)	0.5	(5)
150	12	(40)	2.0	(21)	1.0	(10)
300	24	(80)	4.0	(43)	1.7	(20)
500	40	(133)	6.5	(71)	2.8	(33)
1000	80	(267)	13.0	(143)	5.6	(66)

Over the wire mesh screen place a layer, 100–150 mm (4–6 in) thick, of well-drained compost suitable for rooting cuttings and sowing seed. Alternatively, if plants are to be propagated in trays and pots, a 50-mm (2-in) layer of sharp sand or granulated peat will do.

Above: A simple heated garden frame, using soil-warming cable to provide bottom heat, will serve almost all the functions of a small cool greenhouse. Note that the thermostat is positioned at root level. A hessian mat gives extra heat insulation during winter nights, and the wire mesh is added to protect the heating cable.

As with bench warming in the greenhouse, the temperature of the rooting medium can be controlled with a thermostat. This should be of the rod type, designed for outdoor use, and should be positioned at about root level.

Space or air heating, to give extra protection against frosty weather, can be installed. This can be operated by hand and switched on when frost is predicted. For extra insulation in winter, keep at hand a hessian frame mat, an old piece of carpet, or some sacking, with which to cover the lights at night. Although the inside of the frame can be lined with polythene to give a double glazing effect, this is not always a wholly successful method. It can cause excessive condensation, which may result in disease, and also reduces light, which will limit the growth of plants in winter, even when there is sufficient heat available.

Figure labels: Frame light; Matting; Thermostat; Plants rooting direct in compost; Side of frame; Soil-warming cables; Sharp sand; Wire mesh; Peat and sand mix; Seedlings in box

Electricity

The greenhouse is a hazardous place for electricity. At best it is always slightly damp and at worst will be splashed with water everywhere in the summer in an attempt to keep the temperature down. So, no matter how tempted you may be to bodge wiring jobs—*don't!*

A greenhouse must have its own separate power supply from the meter point in your house, and be fitted with a 20–30 amp mains switch and fuse unit. Then, if anything should go wrong, it is the fuse and not you that takes the shock. Inside the greenhouse the power supply should run either to a properly insulated socket unit, or to a fused mains switch from which spurs are taken to plug sockets and circuit wiring. Needless to say, all fittings *must* be waterproof. From your house to the greenhouse, cable can be run either overhead or underground. The latter way makes less of an eyesore and is probably safer.

Below: Purpose-designed control panel securely fitted to the wall of a greenhouse. Each plug has its own socket: multiple adaptors must never be used.

Underground cables
Three types of cable can be used underground.

4 mm² PVC-insulated and sheathed twin with earth cable. This is very similar to modern cable used in the home but is much heavier. Below ground it must be run through galvanized steel conduit. Rubber grommets are fitted at the ends of the conduit to prevent the cable being damaged.

Mineral insulated copper clad cable. Here two insulated wires—the live and neutral—are run within a copper tube and insulated by a white chalk-like substance. The copper casing acts as an earth. When cut, the cable ends must be sealed against damp. As this is not an easy job it is better to buy cables ready sealed in the required lengths.

Armoured PVC-sheathed cable contains two insulated cables inside a tough PVC sheath bound with steel mesh. Earthing is either by means of the steel casing or a third cable.

Cable must be laid 500 mm (20 in) or more below the surface and in as straight a path as possible from the meter point to the greenhouse. If it can be run under paving, where there is no chance of disturbance, all the better.

Avoid running cable through a flower border unless you are prepared to bury it very far down. All conduit and cable must be firmly fixed to walls, and conduit pipe must be earthed with 2.5-mm² PVC-covered wire.

Overhead cables
For running a wire overhead, be sure to use 4-mm² PVC-insulated and sheathed twin with earth cable. This is suspended at least 3.5 m (11 ft) from the ground and supported by a gantry wire. The wire is fastened to insulators fixed against the outside house wall, where the power cable emerges, and to the greenhouse or a well-braced pole standing next to it. The power cable should pass through the house wall in conduit, be fitted to the gantry wire with insulating tape and buckle cable clips, and run in one length without joins. The gantry wire and power cable must be earthed.

Earthing of overhead and underground cables is essential. Cables are taken either to earth conductors in the home or to an earth leakage circuit breaker, an automatic switching device which cuts the current when a fault occurs in the earth.

Control panels and fittings
Obviously, all fittings must be purpose-designed and securely fitted to the greenhouse frame in a position where they are least likely to be splashed. It should be emphasized that all plugs are fused and multiple adaptors must never be used. Nor should cables be left trailing about in the greenhouse; fix them to the frame wherever possible.

NOTE: Only the briefest details on installing electricity in a greenhouse are given here. Unless you are a competent electrician, all work should be carried out by a professional to the highest possible standards. If you do carry out the installation, but are not an electrician, have the work inspected by your local electricity board before switching on.

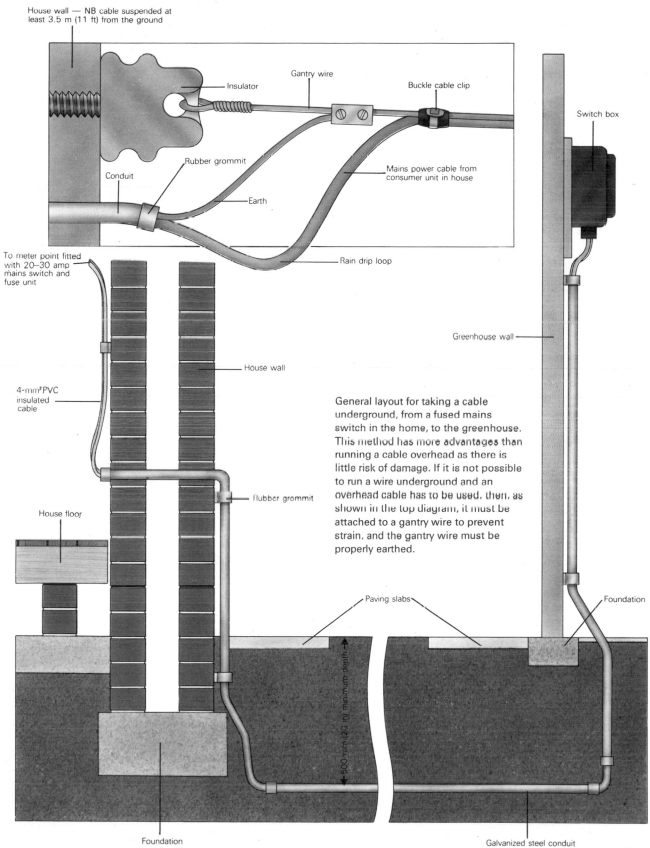

House wall — NB cable suspended at least 3.5 m (11 ft) from the ground

Insulator

Gantry wire

Buckle cable clip

Switch box

Rubber grommit

Conduit

Earth

Mains power cable from consumer unit in house

To meter point fitted with 20–30 amp mains switch and fuse unit

Rain drip loop

House wall

Greenhouse wall

4-mm² PVC insulated cable

House floor

Rubber grommit

General layout for taking a cable underground, from a fused mains switch in the home, to the greenhouse. This method has more advantages than running a cable overhead as there is little risk of damage. If it is not possible to run a wire underground and an overhead cable has to be used, then, as shown in the top diagram, it must be attached to a gantry wire to prevent strain, and the gantry wire must be properly earthed.

Paving slabs

Foundation

500 mm (23 in) minimum depth

Foundation

Galvanized steel conduit

Composts and pots

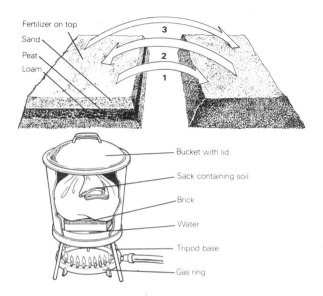

Fertilizer on top
Sand
Peat
Loam

Bucket with lid

Sack containing soil

Brick

Water

Tripod base

Gas ring

Left, below: Arrangement for sterilizing small quantities of soil. The temperature must reach 86°C (180°F) or slightly higher and be maintained for 5 to 10 minutes. The soil must not be placed in the water—it is steamed, not boiled. A large saucepan could be used instead of the bucket.

Seed compost
2 parts sterilized loam
1 part peat
1 part sand

all by volume

42 g (1½ oz) superphosphate of lime
21 g (¾ oz) ground chalk

per 36 l (8 gals) of mix

Basic mix for all potting composts
7 parts sterilized loam
3 parts peat
2 parts sand

all by volume

Fertilizer for No. 1 potting compost
113 g (4 oz) John Innes base fertilizer
21 g (¾ oz) ground chalk

per 36 l (8 gals) of basic mix

Fertilizer for No. 2 potting compost
226 g (8 oz) John Innes base fertilizer
42 g (1½ oz) ground chalk

per 36 l (8 gals) of basic mix

Fertilizer for No. 3 potting compost
340 g (12 oz) John Innes base fertilizer
63 g (2¼ oz) ground chalk

per 36 l (8 gals) of basic mix

Using soil straight from the garden in seed trays and pots leads to countless problems for young plants—inadequate aeration, poor drainage, starvation, stagnation and possibly a host of pests and diseases.

The ideal compost for raising plants within a container should:
1 Have an open structure which allows good aeration and drainage but retains moisture without frequent watering
2 Not compact with frequent watering or shrink when dry
3 Give adequate support to cuttings until they have rooted and hold seeds in position while germination takes place
4 Be reasonably sterile. That is, free from weeds, pests and diseases, and any noxious substances
5 Contain sufficient food to promote healthy growth.

Composts for propagation can be divided into two types: soil-based and soilless. The standard soil-based mixes are the John Innes composts and the proprietary soilless ones consist of just peat or peat and sand.

John Innes composts

John Innes mixtures divide into seed composts and potting composts. Both have strictly defined contents, the ingredients being

Above, top: To mix John Innes compost thoroughly, stack a thin layer of loam at the base, followed by peat and sand, with fertilizer on top. Turn the stack at least three times.

loam, peat and grit-type sand. The loam provides support, retains nutrients and supplies all the minor elements essential for healthy plant growth. Peat is responsible for holding moisture and for aeration, which prevents stagnation. The sand contributes to drainage. Chalk, to prevent acidity, and fertilizers, to provide the major nutrients for healthy, vigorous growth, are also added.

The seed compost can be used for rooting cuttings. It is very well drained and does not contain much fertilizer. During the early part of its life a seedling contains most of the nutrients to sustain itself.

There are three grades of potting compost, each containing a different concentration of fertilizer. The No. 1 grade is used for pricking out seedlings and potting on newly-rooted cuttings; No. 2 for potting on young plants making rapid growth; and No. 3 for gross feeders and potting large plants. For greenhouse growing and propagation the No. 1 and No. 2 grades only are needed.

Buying. John Innes compost should be as fresh as possible when bought. Preferably buy from a retailer who date-stamps his consignments. For, odd though it may seem, John Innes composts are perishable. Six weeks or so after mixing the fertilizers start to break down, releasing ammonia fumes which are harmful to plants, particularly seedlings. This can be prevented with reasonably fresh compost by opening the bags as soon as you get them home, so allowing the fumes to escape.

Mixing your own. The main snags in making your own compost are the difficulties of obtaining small quantities of good quality loam, and sterilizing. But if you want to try, a simple sterilizer and the correct mixing procedure are shown opposite.

Soilless composts

Soilless composts are lighter and cleaner than those which are soil-based. They are also naturally sterile. The main disadvantages concern watering—very dry compost can be difficult to re-wet and over-watering tends to saturate—and anchorage of large plants.

Buying. Proprietary mixes are either pure peat or peat and sand (the latter tends to provide better support for large plants). Composts are normally available in two grades, one for seed sowing and striking cuttings and the other for potting. The potting grade will sustain growth for about six weeks, after which supplementary feeding is necessary. If you want to manage with the minimum number of composts, an all-purpose mix can be bought which is suitable for seed sowing, rooting cuttings and potting. This gives excellent results, but when used for potting, supplementary feeding with a liquid fertilizer is necessary after two or three weeks.

Mixing your own. Special fertilizer packs can be bought to mix in peat for making soilless composts, but four points must be remembered.

1 Peat tends to be very acid: sphagnum peat, which is normally recommended, has a pH of 3.5 to 4.0 (pH is a measure of soil acidity). This must be corrected to 5.0 to 5.5 by adding ground limestone. As a rule 112 g (4 oz) added to every 36 l (8 gals) of peat will raise the pH by a factor of 1.

2 The peat must be fibrous and passed through a 6-mm (¼-in) sieve. Dusty, dirty, greasy or very coarse peat must not be used.

3 A small quantity of fertilizer must be mixed with a large bulk of peat. Spread out the peat in a thin layer, sprinkle the fertilizer evenly over it and then turn the mixture at least three times. The peat should be only slightly damp, otherwise the fertilizer tends to go lumpy and will not mix.

4 Sand, if it is added, should be a coarse horticultural grade reasonably free of lime. Builders' sharp sand is not suitable.

Growth flotation

The principle of growth flotation is very simple; a layer of specially-treated perlite is floated on water, seed is sown on the perlite, covered and allowed to germinate. The perlite acts as compost and seedlings root through it into the water beneath. If a fairly thick layer of perlite is floated on water, cuttings can be rooted by this method. Its advantages are that no checks in growth occur from drying out, and the growing medium used is completely sterile. However, it is a complicated system to set up. It also encourages very fleshy root growth on seedlings and cuttings which can then be damaged during potting in conventional compost. Growth flotation is therefore best left as an interesting technique for the experimenter to try out.

Above: Growth flotation—the seed tray before sowing. The tray is divided into compartments.
Left: A layer of perlite is placed in each compartment and floated on water. Seed is space-sown in each compartment and then covered by a further layer of perlite.
Right: A selection of vegetables after they have fully germinated. The dark material appearing on the perlite is powdered bark which is used to prevent algae from growing on the surface.

Fluid sowing

This technique has been developed to give seed-raised plants a head start by sowing them in the ground after they have germinated. Seed is first sown on moist paper tissues in a closed container, such as a sandwich box, placed in an airing cupboard or similar warm place. When seedling roots start to emerge, the germinated seeds can be transferred to their permanent growing position by means of fluid sowing. The germinating seeds are harvested from the tissue by gentle washing and sieving. They are then carefully stirred into an inert (e.g. wallpaper) paste. The mix of seed and paste is spooned into a polythene bag and the neck of the bag sealed. A small hole is then made in one corner of the bag; seed and paste are squeezed out, like toothpaste from a tube, into a prepared seed drill. After sowing the drill is covered with soil and the germinating seeds allowed to develop in the normal way. Fluid sowing provides a method of starting seeds that are difficult to germinate, such as summer lettuce and parsley, under almost ideal conditions before sowing in open ground.

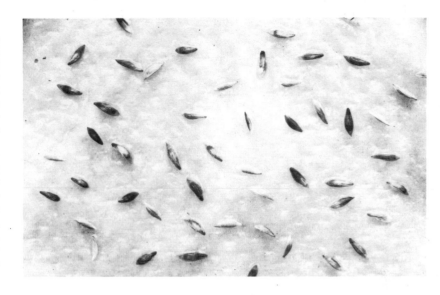

Above: Fluid sowing. Seed is sown and germinated on moist paper tissue inside a sandwich box, or other similar container, which is then put in a warm place, such as an airing cupboard.
Right: The germinated seed is harvested by gently washing off the paper tissue and then passing through a sieve.
Below right: After harvesting, the germinated seed is mixed with paste and spooned into a polythene bag. Paste and seed are squeezed through a small hole in one corner of the bag into a prepared drill.

Left: Pots and trays for propagation.
1 Plastic propagating top for standard sized seed trays
2 Top for half-sized trays
3 Top for pots (various sizes)
4 Full-sized seed trays
5 Half-sized seed trays
6, 8 Plastic pots
7 Fibre pots ('Long Toms')
9 Compressed peat pots
10 Peat strip pots
11 Polystyrene slab pots
12, 13 Jiffy 7s
14 Soil block maker
15 Soil blocks

Fibre and compressed peat pots can be used only once. Peat strip pots are ideal for rooting cuttings while polystyrene slabs are useful in cold greenhouses, insulating compost and roots.

Other compost materials and mixes

There are other types of material suitable for various kinds of propagation. The following are among the most useful.

Cutting compost. Most cuttings must be rooted in a well-drained mix; John Innes seedling compost and its soilless equivalent tend to hold too much water. A far better alternative is a compost composed of equal parts horticultural sharp sand and reasonably fine granulated peat. No fertilizer is added and cuttings are potted as soon as they have rooted.

Sphagnum moss is formed from the dead remains and living portions of an acid bog moss. It is relatively sterile and contains a natural fungicide which inhibits many disease-carrying organisms. Sphagnum is an ideal material for air layering and, when shredded, is a useful medium on which to germinate very fine seed.

Leaf mould can be used as a substitute for peat, but the quality varies according to the types of leaves from which it is made. Oak, elm and sycamore provide the best. Decomposition of the leaves can take a year or more, and it is essential to sterilize the material before use.

Perlite is made from heating a type of volcanic ash; it is a white siliceous material with a light, spongy texture. Particles vary in size from 2 to 3 mm ($\frac{1}{16}$–$\frac{1}{8}$ in). Perlite is completely sterile, very clean to use and is uniform in texture. It drains freely but also retains water well and is chemically inert, so that any fertilizer can be added in precise amounts without creating problems. Common uses for perlite are as a substitute for sand, which tends to be limey, in John Innes composts, and as an ingredient in cutting compost (i.e. 1 part perlite; 1 part peat). It is also used alone as a rooting medium for cuttings.

Vermiculite is similar to perlite, but differs in being not quite so chemically inert. It also tends to

break down with age, causing some deterioration in drainage and aeration when used for composts. There are builders' and horticultural grades of vermiculite—always buy the latter.

Solid block rooting mediums combine compost and pot in one unit.

Soil blocks are cubes of compressed peat-based compost measuring about 50 mm (2 in) across each face. One block acts as a complete rooting medium and pot for a single plant, which can be introduced as a seed, seedling or cutting. The plant can be grown in this until it is ready for planting or potting on. Blocks need careful handling until they become root-bound or they will crumble. Composts used for soil blocks tend to be low in nutrients, so that plants will need feeding with a liquid fertilizer three to four weeks after they have become established.

The advantages of using blocks are that no pots are required unless plants are potted on, there is minimal root disturbance, and no

Above: Plants raised from F_1 seed, such as these begonias growing in soil blocks, are generally more vigorous and uniform in shape and size than normal seed varieties.

problems of waterlogging or aeration arise. Disadvantages include the fact that blocks dry out quickly unless they are stood on capillary matting or a similar automatic watering system, and that all surfaces provide ideal growing conditions for moss and algae.

Proprietary block makers and special blocking composts are also available.

Jiffy 7s are compressed discs of peat, with some added nutrients, encased in fine plastic netting. Standing them in water causes the discs to expand into cylindrical peat-filled blocks. The compost, after wetting and expansion, is soft enough to insert cuttings, seedlings or seeds. The advantages, drawbacks and requirements of Jiffy 7s are much the same as for soil blocks.

Seed propagation

A seed is an embryo plant packed within a hard, protective coat to keep it from being damaged until it has started to germinate. This embryo contains all the potential features of the adult plant—leaves, stem and roots. In addition, it may produce special seed leaves called cotyledons which can act as a food store and, consequently, are thick and fleshy. Normally one or two seed leaves are present. Alternatively, food is stored within special tissue inside the seed. This food store keeps the embryo plant alive until it has germinated and started to manufacture its own food.

Buying seed

You will probably not be able to buy all the different types of seed you require from one firm. Collect a good selection of catalogues by late December, allowing plenty of time to place orders for the varieties you want. Avoid the temptation to order more than you need.

Packaging

Seed is sold either packed in paper envelopes or in air-tight hermetically-sealed foil packets. With the latter, seeds are packed under conditions of low humidity and moisture content and this greatly extends their life. For example, in an ordinary paper packet the viability of onion seed deteriorates rapidly after a year, whereas in a hermetically-sealed packet it keeps from four to five years.

Pelleted seed

Making small seeds into pellets by covering them in a coat of clay-like material enables precise sowing and easier handling, greatly reducing waste. Pellets are very much more expensive than naked seed. They must be kept bone dry until sown, after which it is necessary to keep the pelleting material constantly moist to ensure germination. Most failures result from soil being too dry.

Viability and storage

Always buy freshly packed seed. If in doubt, check the date of packaging.

Seed can be kept for varying lengths of time, depending on the kind and the way it is stored. For instance, under ordinary conditions onion and candytuft lose their viability in less than a year, while acacia seed, with its tough, watertight coat, will remain viable for 20 years or more. To keep seed fresh:

1 Open packets only when you are ready to sow

2 As soon as enough seed has been removed from a packet, reseal it with sticky tape

3 Store all packets—opened and closed—in a dry, moisture-proof container kept in a cool, frost-free place

4 Or put seed packets in a plastic bag, suck out the air, seal with a twist tie and then store in the vegetable compartment of a refrigerator.

The table below gives the storage life of common vegetable and flower seeds. If kept in a refrigerator the storage life will be at least doubled.

F_1 seed

The prefix F_1 indicates a very special type of flower or vegetable seed. Plants raised from F_1 seed are more vigorous—vegetables often produce double or triple the yields of normal varieties—and uniform in size, shape and flowering time. The raising of F_1 seed involves a sequence of careful hand-

Seed storage			
		Aster	2 years
Vegetables		Azalea	1 year
		Begonia	1 year
Asparagus	3 years	Calendula	5 years
Aubergine	4 years	Candytuft	1 year
(egg plant)		Calliopsis	5 years
Bean, broad	3 years	Canna	5 years
Bean, French	3 years	Centaurea	5 years
Bean, runner	3 years	Cineraria	2 years
Beetroot	2 years	Chrysanthemum	5 years
Broccoli	4 years	Cosmea	5 years
Brussels sprout	4 years	Dahlia	3 years
Cabbage	4 years	Delphinium	2 years
Calabrese	4 years	Dianthus	5 years
Capsicum (pepper)	2 years	Eschscholzia	5 years
Carrot	2 years	Ferns	1 year max.,
Cauliflower	4 years		often days
Celery	1 year	Fuchsia	1 year
Chinese	4 years	Helichrysum	2 years
cabbage		Kochia	2 years
Courgette	4 years	Lily	2 years
Cucumber	5 years	Malva	1 year
Kale	4 years	Morning glory	5 years
Leek	1 year	Nicotiana	3 years
Lettuce	3 years	Ornamental	4 years
Marrow	4 years	cabbage	
Onion	1 year	Ornamental	2 years
Parsley	1 year	sweet corn	
Parsnip	1 year	Ornamental	2 years
Pea	3 years	pepper	
Pumpkin	4 years	Papaver	5 years
Radish	4 years	Petunia	5 years
Seakale beet	2 years	Phlox	3 years
Spinach	4 years	Primula	3 years
Spring onion	1 year	Salpiglossis	5 years
Swede	4 years	Saintpaulia	1 year
Sweet corn	2 years	Schizanthus	5 years
Tomato	3 years	Stock	5 years
Turnip	4 years	Tagetes	5 years
		Tropaeolum	5 years
Flowers		Verbena	5 years
		Viola	5 years
Alyssum	5 years	Zinnia	3 years

pollinating operations, often over many years. The object is to fix, by a programme of in-breeding, required characteristics in two separate lines of plants. When the breeder has reached his aims, the two lines are crossed to produce F_1 seed. The resulting offspring not only possess all the desired characteristics of both parents, but they are also exceptionally vigorous. Hence the term 'hybrid vigour' which often accompanies descriptions of F_1 varieties in seed catalogues. Unfortunately this vigour will not show itself in offspring of seed saved from F_1 plants; in fact they will be a very mixed crowd.

Varieties of F_1 vegetable seeds are available for aubergine, broccoli, brussels sprout, cabbage, capsicum, carrot, cauliflower, Chinese cabbage, cucumber, curly kale, marrow, melon, onion, savoy, sweet corn, tomato and turnip. Remember, however, that the cropping season of an F_1 variety can be very short. The entire crop may then mature almost simultaneously, resulting in a glut. On the other hand a normal seed type, because of its inbred variability, will crop over a much longer period. F_1 flower seeds are represented by 30 different types of plants and over 250 varieties.

Below: Germination of the broad bean. This is a hypogeal seed—one where the seed leaves (cotyledons) remain beneath the surface. The cotyledons are a food store, providing the young plant with all the nourishment it needs until the first true leaves develop. The seed root is long, thick and fleshy, making it difficult to transplant without damage.

Germination

Germination will not occur until the seed is provided with the right conditions to induce growth. These are warmth, moisture, an ample supply of oxygen, and, in some cases, light. For this reason, a heated greenhouse or propagator is the ideal environment for the majority of seeds.

The mode of development from embryo to seedling shows great variety. Broadly, the seed root (called a *radicle*) emerges first, grows down into the soil and starts supplying the developing seedling with water and some nutrients. The radicle can be a single thick root (tap root) or a mass of fibrous roots, depending on the type of seed. Next, the seedling shoot (the *plumule*) emerges. Cotyledons may remain below ground (hypogeal germination) or rise above it (epigeal germination), forming the first leaves.

The advantages of raising plants from seed in preference to other methods of propagation are: variety; health and vigour; cost; space.

Variety. Every seed has the potential to give rise to an entirely new plant that will have all the main characteristics of the parent, but differ in a number of details. For example, it may have a slightly different flower colour or branch more freely. It is such minor variations that, through careful selection, give rise to new strains of plants. Wide variability in selected types is avoided by removal of 'rogues', by careful

breeding programmes and the use of F_1 seed.

Health. Seed-raised plants tend to be healthier and more vigorous than plants raised from cuttings, or by grafting, layering, etc. This is because new plants are normally raised from seed each year, and it is not necessary to maintain stocks of old plants as sources of propagating material.

Cost. This is the most obvious advantage: a single seed costs only a minute fraction of the price of buying a plant.

Space. Raising new plants from seed each year means that greenhouse space is not required for overwintering stock to provide cutting material the following season.

Collecting seed

Saving seed from most annual and biennial plants is not usually worth the trouble. The result will normally be a rather mixed bunch of offspring bearing little or no

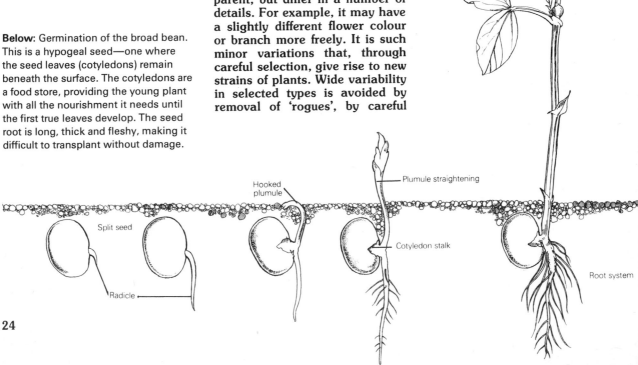

First leaves

Hooked plumule

Plumule straightening

Cotyledon stalk

Split seed

Radicle

Root system

resemblance to the parents. However, it is worthwhile with some herbaceous perennials, trees and shrubs. Seed is collected as it is ready for dispersal from the plant, cleaned and stored in paper bags in a cool, dry place. Seeds from dry heads are shaken or rubbed out, and the chaff removed as they are poured back and forth from one sheet of paper to another, letting the wind blow the lighter pieces away. Large pieces of debris can be picked out with tweezers. Seed enclosed in fleshy fruits and berries is removed by pulverizing the flesh and carefully floating off the pulp in water. Alternatively, soak the seed in water in a warm place for a few days; the flesh will normally start to ferment and float off. After collecting, seed is thoroughly dried before being put into bags. Ensure that each batch of seed is clearly labelled with the name of the plant and the date collected.

Fern spores

Ferns reproduce by spores, not seeds, and have a complex and unusual life cycle. Spores develop on the fern leaves (termed *fronds*) and are discharged during mid- to late summer. Except when massed together on a clean piece of paper, they are invisible to the naked eye. With moist, warm and humid conditions each spore develops into a small, heart-shaped structure, about the size of a pinhead, called a *prothallus*. It is from this, after a sexual phase, that the new fern grows. Fern spores are rarely worth buying as they remain viable for only a very short period, often a matter of days, after dispersal.

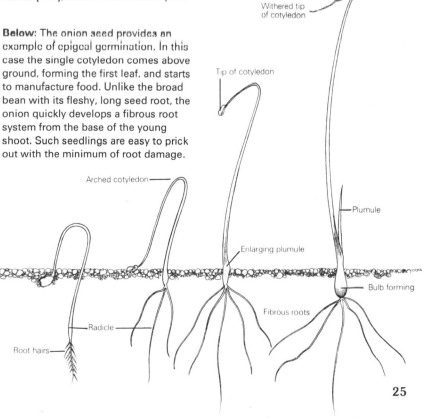

Above: Ferns have a germination cycle which is completely different from that of normal flowering plants. Microscopic spores (**4**) are produced by the millions in pustule like structures. These are normally found on the underside of leaves (**1–3**), called fronds. Each spore germinates to produce a pin-head sized structure called a prothallus (**5**). Germination will normally only occur under warm, very moist conditions. The new fern (**6**) grows from the prothallus after a complex sexual phase.

Below: The onion seed provides an example of epigeal germination. In this case the single cotyledon comes above ground, forming the first leaf, and starts to manufacture food. Unlike the broad bean with its fleshy, long seed root, the onion quickly develops a fibrous root system from the base of the young shoot. Such seedlings are easy to prick out with the minimum of root damage.

Withered tip of cotyledon

Tip of cotyledon

Arched cotyledon

Plumule

Enlarging plumule

Seed

Seed

Bulb forming

Radicle

Radicle

Fibrous roots

Root hairs

Dormancy

Seeds of most annuals, biennials and vegetables grow under a wide range of conditions without any problems, but seeds of many perennials, trees and shrubs are not as simple to germinate. This inactivity in seed is called dormancy, and is a safeguard against unfavourable environmental conditions. The two most common types of dormancy are the result of:

1 A hard seed coat

2 The need for autumn-maturing seed to go through a long cold period before germination can occur. This prevents growth during a warm autumn or winter spell.

Hard seed coats

A hard seed coat can prevent germination by keeping out air and water, by preventing the embryo plant from physically breaking out, or by containing chemical inhibitors to growth. To start germination it is normally only necessary to make a small break in the seed coat by nicking it with the point of a knife or nail file, a process called chitting. Alternatively, the seed coat can be softened by soaking in warm water for a couple of days. This will also remove chemical inhibitors. Seeds with hard coats include sweet pea, dogwood, honey locust and tree peony.

Cold treatment

Gardeners have developed a technique called stratification for cold treatment of seed. This exposes seed to low temperatures of 2–5°C (35–40°F) under moist, airy conditions. This can be achieved out-of-doors or in a refrigerator. Out-of-doors a simple box is constructed, as shown in the diagram, and the seed sandwiched between layers of sharp sand, peat, vermiculite or perlite. The box is set in a frost-free place—the seed must not freeze or be allowed to dry out. Wire mesh stops damage from mice. By spring germination may have started, so care must be taken both when separating the seed from the stratification medium and with sowing. For stratification in a refrigerator, mix the seed in moist perlite or vermiculite, put into a plastic bag, seal and place in the vegetable compartment. Add a little fungicide to prevent any disease organisms developing under the moist, humid storage conditions in which the seeds are kept.

Seeds of some plants require more complex stratification and may need a period of warmth followed by cold—e.g. yew and hawthorn—or cold followed by warmth and then another period of cold—e.g. lily-of-the-valley and solomon's seal—before the shoot will start to emerge.

Seed sowing in containers

Success in raising plants from seed depends very much on using the correct sowing technique. The way a seed is sown will be dictated by its size.

Very fine, dust-like seed. The quantity of such seed in a packet is minute. Mix it with a little dry silver sand to increase the bulk and carefully dust on to the surface of the compost. Do not cover the seed, just lightly firm. Plants producing very fine seed include:

Antirrhinum	Gloxinia
Auricula	Impatiens
Azalea	Lobelia
Begonia	Meconopsis
Calceolaria	Mimulus
Campanula	Polyanthus
Ferns	Primula
Gentiana	Saintpaulia
Gesneria	

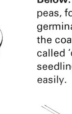

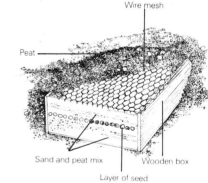

Wire mesh

Peat

Sand and peat mix

Wooden box

Layer of seed

Above: An arrangement set up to stratify seed. This is constructed out in the open and is left over winter to break seed dormancy. A layer of wire mesh stops mice from eating the seed, while the wooden box is packed in peat to keep out frost.

Above: One aid to germination involves soaking seeds which have hard coats. If placed in a bowl of warm water for a day or two, the hard coat will soften. This process also helps to remove any chemical inhibitors which might hinder germination.

Below: Seeds with hard coats (sweet peas, for example) can be helped to germinate by making a small crack in the coat with a sharp knife (a process called 'chitting'). This will allow the seedling to break out of its cover more easily.

Preparing containers for sowing

1
Below: Using soil-based compost.
The compost should be warm, not ice cold. When using plastic trays and pots, it is usually unnecessary to put a layer of roughage in the base to prevent compost being washed out through the drainage holes. Fill the container, ensuring that there are no air pockets.

2
Strike the soil off so that it becomes level with the rim of the container. The surface of the compost should be flat, with no indentations. If there are any large clods of soil (some may be present in low-grade compost) or stones, pick them out.

3
Lightly firm the compost around the edge of the container, using the tips of the fingers. Do not compress the compost too hard: it should not be so solid that it loses its structure. This will impede drainage and in turn cause waterlogging which encourages disease and poor germination.

4
Above: The surface should be as even as possible. Level by lightly firming, applying an even pressure, with a large piece of flat wood. The finished level of the compost needs to be about 10 mm (½ in) below the rim of the container to allow room for watering.

5
Using soilless compost.
Soilless compost (peat, peat and sand, etc.) needs levelling only. Fill the container with compost and strike off any excess so that it is level with the container's rim. Lightly firm the edges and corners with the finger tips.

6
Tap the sides of the container, using the flat of the hand, so as to level the compost. No further preparation is necessary, as settlement takes place with watering.

An hour before sowing, thoroughly soak the prepared containers, watering with a can fitted with a fine rose, and allow the excess water to drain away.

1
Very fine dust-like seed, such as the begonia shown here, is most easily sown when first mixed with a little silver sand to increase its bulk. It is then shaken on to the compost's surface. The seed is not covered.

2
Fine seed (e.g. alyssum) can be sown directly from the packet. Broadcast it thinly over the surface of the prepared compost, and then cover the seed with about 3 mm (⅛ in) of fine, well-sifted compost.

3
Small, flat seed (e.g. lettuce) or round seed (e.g. cabbage) is best sown by taking a pinch between thumb and finger and carefully broadcasting it over the surface. Depth of cover for this type of seed should be 3–6 mm (⅛–¼ in).

Fine seed to cover. This is fractionally larger than very fine seed, and is covered after sowing by a thin layer of compost. It too can be mixed with a little silver sand to increase bulk prior to sowing. But it can also be sown directly from the packet, taking care to broadcast it evenly and thinly over the surface of the compost. Cover the seed with a thin layer—no more than 3 mm (⅛ in)—of finely sifted compost and lightly firm. Seeds sown in this manner include:

Ageratum	Nemophila
Alyssum	Nicotiana
Cacti	Papaver
Celery	Schizanthus
Clarkia	Sedum
Eschscholzia	Streptocarpus
Exacum	Thyme
Linaria	

Small flat and round seed. Seed of this type is sown directly on to the surface of the compost. Sow straight from the packet or take a pinch between thumb and finger and carefully broadcast, avoiding clumps. Expensive seed of F₁ hybrids can be space-sown in rows; allow about 13 mm (½ in) between each seed. After sowing cover with 3–6 mm (⅛–¼ in) of fine compost. Eighty per cent of garden seed falls within this category, the most common being:

Amaranthus	Lettuce
Arabis	Mesembry-
Asparagus	anthemum
Aster	Mustard
Aubergine	Onion
Broccoli	Ornamental
Browallia	cabbage
Calabrese	Parsley
Cauliflower	Penstemon
Centaurea	Pepper
Chives	Phlox
Chrysanthemum	Physalis
Cineraria	Salvia
Cress	Solanum
Dianthus	Stock
Didicus	Tomato
Freesia	Verbena
Fuchsia	Viola
Leek	

Hairy or downy seed. If shaken straight from the packet, this type of seed sticks together, so with finger and thumb separate the individual seeds and space-sow. Otherwise mix the seed with a little sharp sand to separate. Cover with a thin layer of sifted compost and lightly firm. Examples of common hairy seeds are:

Anemone	Gaillardia
Clematis	Marigold
Dandelion	Tagetes

Large seed (up to 6 mm (¼ in) across). This is space-sown in rows, setting seeds 13–25 mm (½–1 in) apart. They may also be placed in small groups or individually in pots. The depth of covering should be two to three times the diameter of the seed. Pelleted seed also falls within this category. Pellets are never sown deeper than their diameter. Large seeds include:

Calendula	Lupin
Canna	Nasturtium
Cosmos	Pelargonium
Cyclamen	Sweet corn
Dahlia	Sweet pea
Euphorbia	Zinnia
Ipomoea	

Very large seed (over 6 mm (¼ in) across). Sow these seeds individually or in pairs in small pots. Covering, depending on the size of the seed, will vary from 13 to 25 mm (½ to 1 in). Flat seeds,

4

Hairy seed (e.g. marigold) tends to stick together in clumps. The simplest method of sowing this is to separate the seeds out by hand before broadcasting.

5

Large seed (e.g. sweet pea) up to 6 mm (¼ in) across is space-sown in rows. It can also be sown in pots, individually or in groups. Depth of cover should be two to three times the diameter of the seed.

6

Very large seeds (e.g. runner bean) should be sown individually in pots so as to lessen root damage when planting out or potting-on. Sow them to a depth of 13–25 mm (½–1 in).

such as melon and cucumber, should preferably be sown on edge. Giving each seed its own pot reduces the amount of root disturbance and damage when it comes to potting on or planting out. Plants which fall into this category include:

Broad bean	Marrow
Castor oil plant	Melon
Cucumber	Pea
French bean	Pumpkin
Gourd	Runner bean

Seedlings

After sowing, clearly label trays and pots, and cover with pieces of glass or plastic propagator tops. This will prevent the compost from drying out, which could kill emerging seedlings. It could also make some seeds, such as lettuce, go dormant, even when subsequently watered. Nor should the compost be kept too wet, as damping off could occur. A precaution against this disease is to water with Cheshunt compound or benomyl. If you have a mist propagation unit, it can be used to germinate seed. Leave sown containers uncovered and remove

from the mist as soon as the first seedlings emerge.

Most seeds will germinate rapidly and evenly at a temperature of 21°C (70°F) or thereabouts. Avoid widely fluctuating temperatures, for these will cause sporadic germination. Some seeds respond better to lower or higher degrees of heat—see the table on pages 34–35.

As soon as seedlings start to appear, remove covers and move to a position where there is good light and a temperature 6–8°C (10–15°F) lower than that used for germination. Low light and high temperatures will cause weak, spindly growth. Keep an eye on watering, for seedlings must not dry out. Standing containers on a capillary mat will solve this problem if you are out at work all day.

Pricking out

Before seedlings become too large and start to compete for light and root space, they must be transplanted. This is called pricking out. Provided seedlings are big enough to pick up, it can be

done when the first true leaves, or cotyledons, have expanded. The danger of root damage is then lessened.

Seedlings can be pricked out into trays or individually into small pots. The latter reduces root disturbance when it becomes necessary to pot them into larger containers or to plant out. Use John Innes potting compost No. 1 or its soilless equivalent. Prepare pots and seed trays as for seed sowing.

Lift a small batch of seedlings with a plant label and carefully tease the individuals apart. Pick up each seedling by a leaf, between finger and thumb. Do not handle the stem as it is easily injured. Using a dibber, make a hole large enough to take the roots comfortably, insert the seedling and lightly firm the compost round the base. Seedlings can be planted at a deeper level than that at which they were originally growing, so there is no danger of them toppling over. If the seedling forms a rosette of leaves, set the lowest leaf at soil level. Very small seedlings can be pricked out in groups of three to

Right: Prick out seedlings from their original pot (top) when they become overcrowded and as soon as they are large enough to handle without damage. Lift each seedling carefully, using a plant label. Prepare a seed tray (centre) and with a dibber make a hole large enough to take the seedling's roots comfortably. Insert the plant and then lightly firm the compost. Space the seedlings so that they have room enough to grow. After pricking out, water well, using a can with a fine rose.

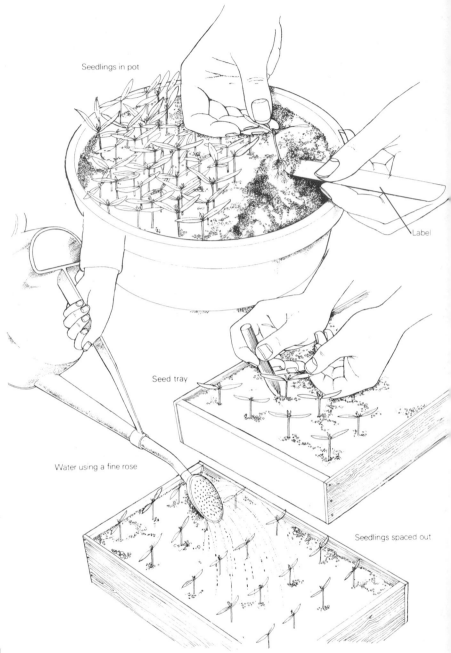

Seedlings in pot

Label

Seed tray

Water using a fine rose

Seedlings spaced out

five if they prove difficult to separate. Never bury the growing point. No more than 40 seedlings should be put in a standard sized seed tray.

Water newly pricked-out seedlings well, using a watering can fitted with a fine rose. Then for two or three days stand them in a warm part of the greenhouse or put them back in a propagator. Avoid draughts and varying temperatures. Heat can be gradually reduced as seedlings make top growth and establish sound root systems. Space plants out as growth demands, to allow ample light and circulation of air.

Pot plants into larger containers when their roots start growing through to the outside of the root ball: use John Innes potting compost No. 2 or its soilless counterpart. A typical potting sequence for a bedding plant would be:

1 Pricked-out seedling (in John Innes seed compost)
2 Move to 60-mm (2½-in) pot (in John Innes potting compost No. 1)
3 Move to 100-mm (4-in) pot (in John Innes potting compost No. 2).

If not potted on, plants will need supplementary feeding six weeks after being pricked out in a soilless compost, and after eight weeks in John Innes compost. Use a liquid tomato fertilizer applied at the recommended rate once a week.

To encourage bushy growth on many bedding plants—e.g. petunia—it may be necessary to pinch out the tip of the leading shoot. Stake and tie plants as necessary.

Hardening off

Never move plants straight from a heated greenhouse into the open air: this could check growth for several weeks. Gradually acclimatize them to cooler growing conditions by first moving them to a cold frame or by turning down the heat in your greenhouse.

An hour or so before planting in the open ground, water plants well. If they have been box-raised, keep as much compost as possible around the roots to lessen damage. If the ground is dry, give it a good soaking at least a day before planting. Try and avoid planting when it is hot and sunny.

Seed sowing in the open ground

The same care must be taken when sowing seed in the open ground as with sowing in pots and

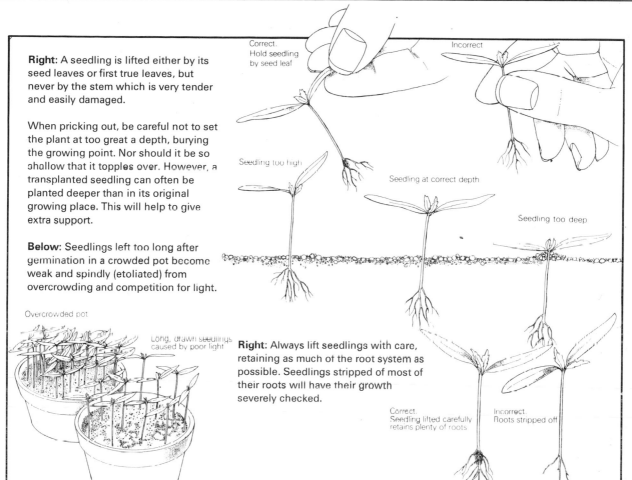

Right: A seedling is lifted either by its seed leaves or first true leaves, but never by the stem which is very tender and easily damaged.

When pricking out, be careful not to set the plant at too great a depth, burying the growing point. Nor should it be so shallow that it topples over. However, a transplanted seedling can often be planted deeper than in its original growing place. This will help to give extra support.

Below: Seedlings left too long after germination in a crowded pot become weak and spindly (etiolated) from overcrowding and competition for light.

Correct. Hold seedling by seed leaf

Incorrect

Seedling too high

Seedling at correct depth

Seedling too deep

Overcrowded pot

Long, drawn seedlings caused by poor light

Right: Always lift seedlings with care, retaining as much of the root system as possible. Seedlings stripped of most of their roots will have their growth severely checked.

Correct. Seedling lifted carefully retains plenty of roots

Incorrect. Roots stripped off

trays inside. Begin with soil preparation.

Ideally, ground should be dug by early winter, giving it time to settle down before final cultivation for seed sowing. By February the soil should be in suitable condition to work into a fine tilth. The first step is to fork it over to a depth of 100 mm (4 in), breaking up large clods of earth and roughly levelling the surface. At the same time any weeds should be cleared out—seedlings dislike competition. Now lightly firm the surface by treading it and then rake to produce a fine, crumb-like structure. Seed needs to be surrounded by fine soil particles (as with seed-sowing compost). There must also be good aeration and an ample supply of moisture. The tilth should be 25 mm (1 in) deep at least, enabling straight, smooth-sided drills to be drawn out. On very light soil poor water-

Right: Potting plants into larger containers should be a gradual process, increasing the pot size by 25–50 mm (1–2 in) each time. A good indicator that a plant needs potting-on is when roots start to grow through to the outside of the root ball.

Potting-on

Seedlings pricked out singly into pots

1

Below: Dig soil during the winter, so that by February it will be ready to work into a fine tilth. First fork soil over to a depth of 100 mm (4 in), breaking up clods.

2

Below: Roughly level and lightly firm the soil by treading. Never work on the ground if it is wet (this causes it to compact) or frozen.

3

Below: Rake over the soil to produce a fine tilth at least 25 mm (1 in) deep. This enables straight, smooth-sided drills to be drawn out.

retaining properties will impede germination. Forking a 25-mm (1-in) thick layer of loam or peat into the top 75–100 mm (3–4 in) of soil will help to improve the situation.

Whatever work you do, make sure first that the soil is in the right condition. The ground must not be frozen, and neither do you want to be up to your knees in mud. Burying frozen soil slows down the warming up in spring and will delay germination. Working on very wet soil will compact the surface, making raking difficult or impossible.

In the early part of the year, a week or two before sowing, cloches lined over the position of drills will help to warm the soil. A couple of days before sowing, work a general fertilizer into the top 50 mm (2 in) at the rate of 140 g per square metre (4 oz per sq yd). Do not sow seed if the soil is very wet. Dry soil can be well watered a day or so before sowing.

The width and depth of a drill will depend on the size of seed and method of sowing. Take out drills against a line to ensure you make straight rows, and to keep them parallel set out the distances between drills with a measuring stick.

Small seed, such as lettuce, is sown in shallow drills 6–9 mm (¼–⅜ in) deep. These can be difficult to take out with a hoe, so use a pointed stick. Deeper drills for larger seeds sown in single rows are taken out with a draw hoe. Shallow trenches for seeds sown in double or triple rows (e.g. broad beans and peas) can be taken out with a spade or draw hoe.

Treat seed sown in the early part of the year with a fungicidal seed dressing to help prevent rotting.

Space-sow seed in drills. Small seed such as lettuce and cabbage is sown about 13 mm (½ in) apart. Middle-sized seed, such as beetroot and seakale beet, is set 18–25 mm (¾–1 in) apart. Large seed, including peas and all types of beans, is sown 25–75 mm (1–3 in) apart in rows. Broadcast sow-

ing—scattering seed—is wasteful. If the soil is a little dry, water drills half-an-hour before sowing small seed. Cover drills by shuffling soil in with your feet or by raking. Level, and lightly firm. Label each row so that you know what has been sown and when.

Thinning out seedlings is a gradual process completed in two or three stages. Where there are gaps in rows, thinnings can be used to fill them. Large seeds should be sown at their final spacing and not need thinning. When the soil is dry, give seedlings a good watering immediately after thinning to settle the disturbed soil back round the roots. Weeds should be kept under control at all times, but it is absolute-

Above: Sow small seed in shallow drills taken out with a pointed stick, using a line to keep the rows straight.

Below: Space-sow seed 13–25 mm (½–1 in) apart, depending upon size. Take a pinch of seed between finger and thumb and drop one seed at a time.

Above: Seed of some vegetables, for example peas, can be sown in double or triple rows. Make a trench about 50 mm (2 in) deep and 150–200 mm (6–8 in) wide with a spade. Again, use a line to keep the trench straight.

Above: Deep drills are made with a draw hoe.

Below: Some seeds (e.g. marrow, sweet corn and cucumber), are sown in groups of three or four at their final spacing. The weaker seedlings are pulled out after germination, leaving the strongest plant.

ly essential to prevent seedlings from being smothered during their early stages of development.

Transplanting—lifting young plants from a seed bed and replanting in a permanent site—is best done in late afternoon to early evening or during dull weather. Water plants thoroughly two hours or so before lifting them carefully with a fork or trowel. Be sure to keep a fair-sized ball of soil round the roots.

When planting, take out a hole large enough to give the roots ample room. Firm well, without ramming soil solidly around the roots, and water well. Ground for transplants should be prepared as for seed sowing, although it is not necessary to obtain such a fine tilth.

Below: Most seed sown outdoors in the open ground will be vegetables. The table below gives a guide to sowing dates, depth of sowing and final spacing.

Vegetable	Date of sowing	Depth of sowing mm (in)	Space between rows mm (in)	Final spacing in rows mm (in)
Asparagus	Mar–Apr	25 (1)	300 (12)	150 (6) transplant to 380 (15) apart in rows set 1200 (48) apart
Artichoke, globe	Apr	20 (¾)	450 (18)	600 (24)
Bean, broad	Nov or Feb–May	50 (2)	600 (24)	50 (2)
Bean, French	Apr–Jun	50 (2)	450 (18)	50 (2)
Bean, runner	May–Jun	50 (2)	900 (36)	75 (3)
Beetroot	Mar–Jun	25 (1)	250 (10)	100 (4)
Broccoli	Apr–May	13 (½)	600 (24) after transplanting	600 (24) after transplanting
Brussels sprout	Mar–May	13 (½)	600 (24) after transplanting	600 (24) after transplanting
Cabbage	Mar–Jul depending on type	13 (½)	450 (18) after transplanting	300–450 (12–18) after transplanting
Cabbage, Chinese	May–Jul	20 (¾)	300 (12)	250 (10)
Carrot	Mar–Jun	13 (½)	250 (10)	Early varieties not thinned Main crop 75 (3)
Cauliflower, winter	May	13 (½)	600 (24) after transplanting	600 (24) after transplanting
Chicory	May–Jun	13 (½)	250 (10)	220 (9)
Cucumber, outdoor	May	50 (2)	600 (24)	450 (18)
Kale	Apr–May	13 (½)	600 (24) after transplanting	600 (24) after transplanting
Leek	Feb–Mar	20 (¾)	300 (12) after transplanting	150 (6) after transplanting
Lettuce	Mar–Aug	13 (½)	300 (12)	200 (8)
Marrow	May	50 (2)	1200 (48)	450 (18)
Onion	Mar–Apr	20 (¾)	300 (12)	125 (5)
Parsnip	Mar–May	13 (½)	250 (10)	200 (8)
Pea	Feb–Jun	50 (2)	900 (36)	50 (2)
Radish	Feb–Aug	13 (½)	150 (6)	25 (1)
Rhubarb	Apr	25 (1)	900 (36) after transplanting	900 (36) after transplanting
Seakale beet	Mar–Apr	25 (1)	380 (15)	200 (8)
Spinach	Apr–May	13 (½)	300 (12)	150 (6)
Swede	Jun	13 (½)	380 (15)	220 (9)
Sweet corn	Apr–May	25 (1)	450 (18)	450 (18)
Turnip	May–Jun	13 (½)	300 (12)	150 (6)

Hardy herbaceous and alpine plants

These are either perennial (living and flowering for several years) or biennial (flowering the year after sowing and then normally being discarded). Seed can be sown in the open, following the procedure set out on pages 32–33, or, in the case of very small seed, in pots. Young plants are transplanted to their permanent growing positions in September or October. A few will need to be stratified—subjected to a period of cold—before they will germinate (see page 26).

Trees and shrubs

Aim to purchase seed between October and December. As soon as it arrives, sow in containers, using John Innes seed compost, and stand outdoors in a sheltered position to avoid freezing. During April bring the containers back into the greenhouse or stand in a heated frame to germinate. Seed collected fresh in the autumn can be stratified and sown in April. For sowing technique see pages 28–29.

Treatment of seed

Vegetables

Name Sowing time	Germination	
	days	temp. °C (°F)
Aubergine (egg plant)		
Feb–Mar	14–21	18 (65)
Capsicum (pepper)		
Feb–Mar	14–21	18 (65)
Celery		
Mar–Apr	10–14	18 (65)
Plant out May–Jun		
Cucumber, greenhouse		
Jan–Mar	14–21	18 (65)
Cucumber, outdoor		
Apr–May	10–14	18 (65)
Plant out Jun		
Lettuce		
Aug–Dec	7–14	15 (60)
Winter–spring cropping		
Marrow		
Apr	10–14	18 (65)
Plant out Jun		
Onion		
Jan	14–21	15 (60)
Plant out Apr		
Sweet corn		
Mar	14–21	12 (55)
Plant out Jun		
Tomato, greenhouse		
Jan–Mar	10–14	21 (70)
Tomato, outdoor		
Mar–Apr	10–14	18 (65)
Plant out Jun		

Spring and summer bedding plants

Name Sowing time	Germination	
	days	temp. °C (°F)
Ageratum		
Mar–Apr	14–21	18 (65)
Plant out May–Jun		
Alyssum		
Mar	7–14	18 (65)
Plant out May–Jun		
Apr–May sow in situ in open ground		
Amaranthus		
Feb–Mar	7–10	18 (65)
Plant out May–Jun		
Apr–May sow in situ in open ground		
Antirrhinum		
Jan–Apr	10–14	18 (65)
Plant out Apr–Jun		
Jun–Sep	10–14	
Overwinter in		
frame with frost protection		
or cool greenhouse		
Aster		
Mar–Apr	14–21	18 (65)
Plant out May–Jun		
Begonia semperflorens		
Begonia, tuberous rooted		
Dec–Feb	14–21	21 (70)
Plant out Jun		
Calceolaria rugosa		
Jan–Mar	10–20	15 (60)
Plant out Jun		
Candytuft		
Mar–Apr	7–10	15 (60)
Plant out May–Jun		
(Iberis)		
Mar–May sow in situ in open ground		
Castor-oil plant (Ricinus)		
Jan–Feb	14–21	21 (70)
Plant out May–Jun		
Celosia (Prince of Wales' Feathers)		
Mar–May	7–14	21 (70)
Plant out from Jun		
Centaurea		
Feb–Mar	21–28	24 (75)
Plant out May		
Chrysanthemum		
Feb–Mar	14–21	15 (60)
Plant out May		
Cineraria maritima		
Mar	14–21	18 (65)
Plant out May		
Cleome (Spider plant)		
Feb–Mar	21–28	18 (65)
Plant out May–Jun		
Coleus		
Feb–Mar	14–21	18 (65)
Plant out May–Jun		

Name	Germination	
Convolvulus		
Mar–Apr	7–14	18 (65)
Plant out May–Jun		
Cosmea (Cosmos)		
Feb–Mar	10–14	15 (60)
Plant out May–Jun		
Dahlia		
Mar–Apr	14–21	21 (70)
Plant out May–Jun		
Daisy (Bellis perennis)		
May–Jun sow in open ground		
Transplant to spring flowering		
position in Sep–Oct		
Dianthus		
Jan–Feb	7–14	15 (60)
Plant out May–Jun		
Eucalyptus (Blue gum tree)		
Feb–Apr	21–35	24 (75)
Plant out May–Jun		
Fuchsia		
Jan–Mar	21–28	24 (75)
Plant out Jun		
Gazania		
Mar	14–21	18 (65)
Plant out May–Jun		
Geranium—see Pelargonium		
Gerbera (Transvaal daisy)		
Feb–Mar	14–21	15 (60)
Plant out May–Jun		
Grevillea (Silk oak)		
Mar–Apr	21–28	21 (70)
Plant out May–Jun		
Heliotropium		
Feb–Mar	21–28	21 (70)
Plant out May–Jun		
Impatiens (Busy Lizzie)		
Mar–Apr	21–28	21 (70)
Plant out Jun		
Kochia (Summer cypress)		
Mar	5–7	21 (70)
Plant out Jun		
Lobelia		
Feb–Mar	10–14	18 (65)
Plant out May–Jun		
Marigold		
Mar–Apr	7–10	18 (65)
Plant out May–Jun		
Mesembryanthemum (Livingstone daisy)		
Feb–Mar	14–21	18 (65)
Plant out May–Jun		
Myosotis		
May–Jun sow in open ground		
Transplant to spring flowering		
position Sep–Oct		
Nemesia		
Mar–Apr	14–21	18 (65)
Jun–Sep	10–14	
Overwinter in cool greenhouse		
but ensure plants continue		
growing		
Nicotiana (Tobacco plant)		
Mar–Apr	7–14	18 (65)
Plant out Jun		
Pansy		
Mar	14–21	18 (65)
Plant out May–Jun		
Jun–Jul sow in open ground		

Transplant to spring flowering
position in Sep–Oct

Perilla
Mar   14–21   21 (70)
Plant out Jun

Petunia
Mar–Apr   10–14   18 (65)
Plant out Jun

Phlox drummondii
Mar–Apr   10–14   15 (60)
Plant out May–Jun

Polyanthus
Mar–Apr   21–28   18 (65)
Plant in spring flowering
position Sep–Oct
May sow in open ground

Primrose—see Polyanthus

Salvia
Mar   14–21   21 (70)
Plant out May–Jun

Schizanthus (Poor man's orchid)
Mar–Apr   14–21   18 (65)
Plant out Jun

Stock
Mar   10–14   15 (60)
Plant out May

Sweet pea
Jan–Feb   14–21   18 (65)
Plant out May
Sep–Oct   14–21 Frame or cool
Plant out May   greenhouse

Tagetes
Mar–Apr   7–14   18 (65)
Plant out May–Jun

Verbena
Mar   21–28   18 (65)
Plant out May–Jun

Viola
Feb   14–21   13 (55)
Plant out May

Wallflower
May–Jun sow in open ground
Transplant to spring flowering
position in Sep–Oct

Zea (Ornamental sweet corn)
Mar–Apr   7–14   18 (65)
Plant out May–Jun

Zinnia
Apr   7–14   18 (65)
Plant out May–Jun

Flowering and foliage house and greenhouse plants

Name Sowing time	Germination	
	days	temp. °C (°F)
Abutilon (Flowering maple)		
Mar–Apr 14–21		21 (70)
Asparagus		
Mar–Apr 21–28		21 (70)
Begonia		
Mar–Apr 14–20		21 (70)
Do not cover seed		
Browallia		
Mar–Jun 14–21		18 (65)
Bryophyllum		
Mar–Apr 21–28		21 (70)

Cactus
Mar–May   14–42   21 (70)
Germination very erratic
Lightly shade seedlings
and young plants

Calceolaria
Jun–Jul   10–20   15 (60)
Flowers spring following
year

Canna (Indian shot plant)
Feb–Mar   21–35   24 (75)
Soak seed in tepid water
for 24 hours before sowing

Capsicum (Ornamental pepper)
Mar   14–21   21 (70)

Celosia (Prince of Wales' Feathers)
Mar–May   7–14   21 (70)

Chrysanthemum
Mar–Apr   14–21   15 (60)
High temperatures tend to
inhibit germination

Cineraria
Apr–Jul   10–14   15 (60)
Flowers Dec–Apr

Cobaea (Cathedral bells)
Feb–Mar   18–21   21 (70)

Coleus
Feb–May   14–21   18 (65)

Cuphaea (Cigar plant)
Feb–Mar   18–21   21 (70)

Cyclamen
Feb–Mar or   21–35   18 (65)
Aug–Sep
Germination tends to be erratic

Cyperus (Umbrella plant)
Mar–May   14–21   15 (60)
Keep compost very moist

Didicus (Blue lace plant)
Mar–Apr   7–10   18 (65)

Eucalyptus   21–35   24 (75)
(Blue gum tree)
Germination can be erratic
and poor

Exacum
Mar–May   14–21   21 (70)

Ferns
Any time   14–35   21 (70)
'Germination' is very erratic
Viability of spores very variable

Freesia
Mar–Apr   28–35   15 (60)

Fuchsia
Mar–May   21–35   24 (75)
Resulting plants very variable
Germination can be erratic

Geranium—see Pelargonium

Gerbera
Mar–May   7–14   10 (50)

Gesneria
Mar–May   14–35   24 (75)
Germination tends to be good

Gloriosa
Jan   21–28   18 (65)

Gloxinia
Jan–Mar   14–21   24 (75)
Do not cover seed

Gossypium (Cotton plant)
Mar–May   14–21   21 (70)

Grevillea (Silk oak)
Mar–Apr   21–28   21 (70)

Hibiscus
Feb–Mar   14–21   24 (75)

Hippeastrum (Amaryllis)
Mar–Apr   10–20   18 (65)
Takes at least 3 years to flower

Hypoestes (Polka dot plant)
Mar–Apr   21–28   21 (70)

Impatiens (Busy Lizzie)
Mar–Apr   21–28   21 (70)

Jacaranda
Feb–Apr   21–28   24 (75)
Germination can be poor

Kalanchoe
Feb–Mar   14–28   15 (60)
Germination is erratic

Lobelia
Mar–Apr   5–10   15 (60)

Mimosa pudica (Sensitive plant)
Mar–Jun   21–35   21 (70)

Musa (Banana)
Mar–May   21–35   21 (70)
Germination is erratic

Palm (many types)
Apr–Jun   1–6   27 (80)
Soak seed   months
in tepid water for
24–48 hours before sowing
Germination is very erratic

Passiflora (Passion flower)
Feb–Mar   21–35   24 (75)
Germination is erratic

Pelargonium ('Geranium')
Feb–May   14–21   21 (70)

Pilea (Moon valley plant)
Any time   14–21   24 (75)

Primula
Mar–Jul   14–21   18 (65)
Sowing time will depend on
type and flowering period

Punica (Pomegranate)
Feb–May   21–42   24 (75)
Germination is erratic

Saintpaulia (African violet)
Mar–Apr   14–28   24 (75)

Schizanthus (Poor man's orchid)
Jul–Sep   14–21   18 (65)

Sinningia—see Gloxinia

Solanum (Christmas cherry)
Feb–Mar   10–14   18 (65)

Sparmannia (African hemp)
Mar–Jun   21–28   21 (70)

Stephanotis (Madagascar jasmine)
Mar–Jun   21–42   27 (80)

Strelitzia (Bird of Paradise flower)
Mar–Apr   3–6   21 (70)
                months
Germination is slow and erratic

Succulent plants—see Cactus

Thunbergia (Black-eyed Susan)
Feb–Apr   21–28   21 (70)

Torenia (Wishbone flower)
Feb–Apr   14–21   21 (70)

Trachelium
Jan–Mar   14–21   21 (70)

Tropaeolum
Feb–Apr   21–28   21 (70)

Propagation by cuttings

A cutting is a portion of plant that carries within it the ability to regenerate all the organs needed to form a completely new plant. But as all plants are supposed to be able to perpetuate themselves by seed, a first reaction may well be that propagating from cuttings is a rather pointless exercise. This is not so; there are a number of valid reasons, the most important being:

1 A plant may not produce viable seed; among the examples of these are figs and many houseplants

2 Cuttings provide a faster and simpler method of propagation; examples include many trees and shrubs

3 Seeds may not breed true and produce offspring identical to the parent plant; among examples of plants such as these are chrysanthemum and carnation varieties.

Selecting material for cuttings

Cuttings should only be taken from healthy, disease-free plants. After preparing stem cuttings, as a precaution against pests dip them into a general insecticide and fungicide such as malathion, plus benomyl. In most cases cuttings can be taken with equal success from flowering and non-flowering growths. However, all flower buds must be removed as these will compete for the food materials the cuttings need to form roots. Hardwood stem cuttings root better when they are taken from the basal portions of shoots, while softwood cuttings root more readily when prepared from shoot tips. Side shoots (lateral growths) tend to root more readily than main (terminal) growths.

Never take cuttings from plants that are wilting or showing signs of starvation. Always select material that is characteristic of the plant. Abnormal growth is generally to be avoided.

Cuttings are grouped according to the part of a plant from which they come. Shoot cuttings consist of a stem or stem plus leaves. Leaf bud cuttings have a small portion of stem containing a single bud and usually a leaf. Leaf cuttings comprise an entire leaf with stalk, or a portion of leaf. Root cuttings are normally small portions of root, lacking stems, leaves and a fibrous root system.

Below: Steps in preparing softwood cuttings. Unprepared cutting (right); cutting trimmed to just below a leaf joint (centre right); and two prepared cuttings (left), with lower leaves removed, ready to insert.

Stem cuttings

Most plants propagated from cuttings are raised from portions of stem. The type of cutting taken will depend on the plant, the condition of the wood and the time of year. Broadly speaking, stem cuttings can be divided into four groups: herbaceous; softwood; semi-ripe or semi-hardwood; and hardwood. For the practical purposes of preparing and rooting cuttings, herbaceous and softwoods can here be considered together.

Herbaceous cuttings, as the name implies, are those taken from herbaceous plants such as lupins, delphiniums, carnations, chrysanthemums and pelargoniums, which rarely produce woody growth. On the other hand, softwood cuttings will, with age, develop into semi-ripe and then hardwood cuttings. A char-acteristic of many herbaceous cuttings is to exude sticky sap. Traditional practice is to let this dry, preventing the entry of disease organisms through the wound, before inserting the cuttings. But, with a few exceptions, it is better not to do this. Drying also results in water loss and wilting, which weakens the cutting and reduces its chances of rooting. Instead, protect the base of the cutting with a fungicide.

Softwood cuttings. These are made from new growths of deciduous and evergreen plants which have not begun to harden. Best results are gained from those taken in the spring. The type of material used will vary from species to species, but avoid soft, thick, sappy shoots; thin, weak ones; and growths that are starting to turn woody. The length of the cutting will also vary according to the plant from which it is taken, but the average will be between 75–125 mm (3–5 in) and have at least two well-developed leaf joints (nodes).

Cuttings are prepared by slicing off the base to just below a leaf joint, and then removing the lower one or two leaves to give a clean length of stem for inserting into the cutting compost. Buried leaves tend to rot, resulting in disease. The base of the cutting should be dipped in a fungicide or rooting hormone preparation. This will prevent rotting and the rooting hormone will also encourage quicker root development.

Softwood cuttings wilt quickly and must be handled speedily to prevent them from drying out. Cuttings can be rooted in pots, boxes or directly in a prepared propagating bench. They need high humidity to prevent wilting and, preferably, a temperature of 21°C (70°F) at the base.

Left: A mallet cutting is one in which an entire section of older wood is retained at the base. Hardwood cuttings of berberis root better when prepared this way.

Right: Normal practice is to prepare a cutting by making the basal cut immediately below a leaf joint. But many semi-ripe and hardwood shrub cuttings will root just as well if the basal cut is made mid-way between two leaf joints.

Left: For a heel cutting, a shoot is carefully pulled off a stem so that it retains a slither of old wood attached to the base.

Right: An internodal cutting is just a short piece of stem which has one or two leaves and buds attached. Although usually used for the propagation of clematis, many plants can be raised from this type of cutting.

Unless being rooted with the aid of a mist propagator, cuttings will also need shading and watering, if possible by regular syringing with tepid water, to prevent scorching and wilting.

If you are working without a propagator, cuttings will root successfully in pots covered with clear plastic bags. Three to five weeks is the time softwood cuttings generally take to root.

Semi-ripe or semi-hardwood cuttings. Semi-ripe cuttings are taken in mid-summer from shoots of the current season's growth which has started to firm. Most broad-leaved evergreens, such as camellia, rhododendron, holly and euonymus, can be raised from this type of cutting.

Cuttings will vary in length from 40 to 150 mm (1½ to 6 in), depending on the plant. They can be prepared from a trimming taken immediately below a leaf (node cutting); a cutting taken mid-way between two leaf joints (inter-nodal cutting); or from a portion of stem, gently pulled off the parent plant, that bears a small piece of old wood attached to the base (heel cutting). Lower leaves are removed to about half-way along the cutting as this gives a clean piece of stem to insert in the cutting compost.

Treating the base of cuttings with a hormone preparation will aid rooting. Semi-ripe cuttings are not as susceptible to rot as softwood ones, but it is still worthwhile treating them with a fungicide.

Semi-ripe cuttings will root in much cooler conditions than will softwoods, even in an unheated frame, but they do best when standing on a heated bench or propagator with a bottom heat of 21°C (70°F). If space is limited, very large leaves can be cut in half, allowing cuttings in the propagator to be packed much closer together.

Cuttings should be rooted and making new growth by the autumn.

Hardwood cuttings can be treated in two ways, depending on whether they are taken from

deciduous plants or conifers.

Nearly all conifers present propagation problems, a fact reflected in the high prices charged for them by nurserymen. The main difficulty is the high resin content of the wood which inhibits root development or makes it a very slow process. But propagation is worth trying and if the following points are kept in mind, an attempt should be reasonably successful.

1 Take cuttings between late autumn and mid-winter.

2 Cuttings from young plants root better than cuttings taken from mature specimens; make them 125–150 mm (5–6 in) long.

3 Heel cuttings are preferable.

4 Remove resin that congeals on the base of cuttings by dipping them for a few minutes in warm water, about 38°C (100°F).

5 Dip the base of cuttings in a highly-concentrated hormone rooting powder.

6 Allow cuttings plenty of light.

7 Root cuttings in a propagator with a bottom heat of 24°C (75°F) or, preferably, on a mist bed.

8 If rooting has not occurred by late spring, wound the base of the cutting and try striking it again.

Above: Plants raised in a greenhouse, from cuttings or seed, should be placed in a cold frame for a week or two before being planted out in the open.

Deciduous hardwood cuttings can easily be rooted outdoors. Taking these is an easy, reliable method of increasing a wide range of hardy trees and shrubs (including roses and some fruit). All you need are a sharp knife or secateurs, a reasonably sunny, sheltered spot and a small cold frame or glass cloches to protect against frost.

Deciduous hardwood cuttings, unlike softwood and semi-ripe cuttings, are not readily perishable. They are prepared at any time during the dormant season, from November to March, and are taken from the previous season's growths. Only shoots that are vigorous, healthy and free from pests and diseases should be used.

The length of cuttings can vary from 100 to 900 mm (4 to 36 in), but the average size is from 150 to 200 mm (6 to 8 in). Each cutting should have at least two nodes from which new shoots have a chance to develop. In theory the basal cut should be made immediately below a node, but the site of the cut makes little difference to rooting, and any portion of stem can be used.

Besides 'straight' cuttings, others can be prepared with a 'heel' or 'mallet'. Including an older piece of wood does give better results with some plants, berberis for instance, especially if hormone treatment is not used.

It is important not to put cuttings in upside-down, as the top end will never develop roots. Give the top a slanting cut to distinguish it from the base, but keep the basal cut straight, as slanted ones are more inclined to rot.

When it comes to inserting cuttings, a light, sandy loam not prone to waterlogging in winter is best. If your soil is on the heavy side, work in peat and sharp sand to a depth of 100 mm (4 in) at

least. For cuttings rooted in containers, John Innes seed compost is very suitable. Cuttings in the open ground must be kept free from weeds, and watering is normally needed in the spring when they start to shoot.

Cuttings can be planted immediately they are prepared. They should be set down two-thirds of their length in a narrow V-shaped trench and firmed securely. Rooting will either take place straight away or simultaneously with shoot growth in the spring. A disadvantage of this method is the possible loss of unrooted cuttings from standing throughout the winter in wet, cold soil.

A more successful technique is to adopt the commercial practice of storing cuttings through the winter and planting them out in early spring. Prepared cuttings, clearly labelled, are tied

into bundles and packed in boxes, using slightly damp peat or sharp sand, before being stored in a cool but frost-free place. Under these conditions, cuttings will start the rooting process, and when planted in spring make rapid root and shoot growth.

Roses raised from hardwood cuttings have no problems with suckers as the root-stock is eliminated. However, hybrid teas and floribunda varieties tend to be slightly less hardy than their budded counterparts. Cuttings 200–300 mm (8–12 in) long, prepared from pruning waste, make bushes in about 18 months.

Below: 1 Preparing hardwood cuttings from dormant, leafless one-year-old shoots. **2** Treating with hormone rooting agent. **3** Prepared cuttings bundled and packed for overwintering. **4** Autumn-prepared cuttings of privet and juniper. **5** Cuttings set to two-thirds their length in a narrow, V-shaped trench. **6** Rooted cuttings.

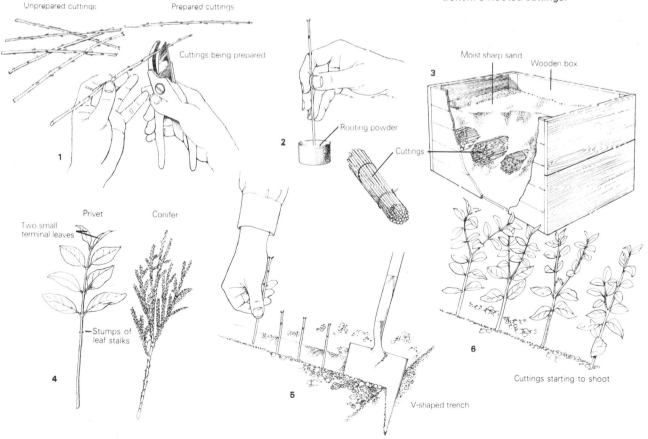

Unprepared cuttings

Prepared cuttings

Cuttings being prepared

Rooting powder

Moist sharp sand

Wooden box

Cuttings

Privet

Conifer

Two small terminal leaves

Stumps of leaf stalks

V-shaped trench

Cuttings starting to shoot

1 2 3 4 5 6

Hardwood cuttings take 6–12 months to develop a good root system. Those rooted in the open are lifted in late autumn after leaf fall. Then, either potted or planted out in a spare piece of ground, they are grown on for a year until they make fair sized plants. For potting, use John Innes potting compost No. 2.

Leaf cuttings

These cuttings are taken from portions of the leaf blade, the whole leaf blade or the leaf blade plus leaf stalk (petiole). Select well-developed leaves which show no signs of ageing or yellowing.

Below left: Propagating anemone plants from root cuttings. **1** Lifting roots. **2** Selecting roots of a good average thickness. **3** Cutting roots into pieces 25–50 mm (1–2 in) long. **4** New shoots developing from portions of roots.

Cuttings are rooted under warm and, in the case of tropical plants, humid conditions.

Sectional leaf cuttings. The leaf blade is cut either into strips 40–50 mm (1½–2 in) wide, e.g. sansevieria and streptocarpus, or portions of 625 mm^2 (1 sq in), e.g. *Begonia rex*. Each piece of leaf must have part of a well-developed vein. Treat the basal edge with a hormone rooting powder and insert cuttings vertically 13 mm (½ in) deep in a well-drained cutting compost. Leaf cuttings will not root if placed upside-down, so if in doubt, mark the basal end with a slanting cut. The new plant develops from the

Below: Propagation of a rhododendron by leaf bud cutting. **1** and **2** Cutting prepared by removing a small shield-shaped piece of stem with a bud and leaf. **3** Roots developing. **4** Root and shoot growth after three to six months.

base of the cutting, and the portion of leaf eventually dies.

Whole leaf cuttings. Streptocarpus and *Begonia rex* are easily raised from whole leaf blade cuttings. Cut the large veins on the underside of a leaf, lay flat on cutting compost and hold it down with a few wire staples. New plants will develop from the severed veins.

Whole leaf cuttings with petiole. For these the leaf is detached together with its petiole, which is inserted in compost, and a new plant develops from the cut base of the petiole. If propagating material is in short supply, the original leaf can be cut off and rooted again to provide a second plant. Saintpaulia (African violet) and peperomia are readily increased in this way.

A fascinating variation of leaf propagation is shown by *Kalanchoe (Bryophyllum) pinnata* and

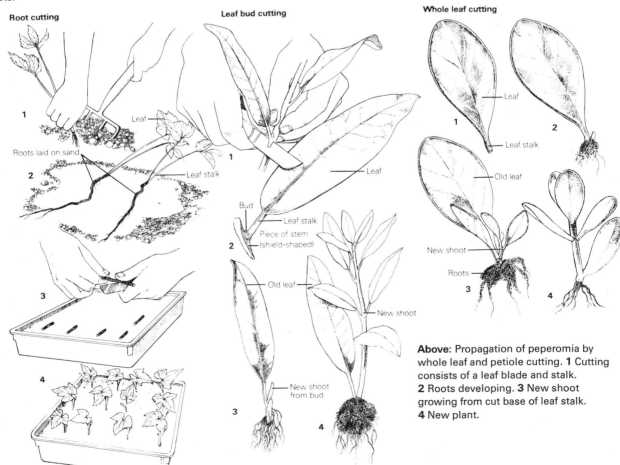

Root cutting

Leaf bud cutting

Whole leaf cutting

Leaf

Roots laid on sand

Leaf stalk

Bud

Leaf stalk

Piece of stem (shield-shaped)

Old leaf

New shoot

New shoot from bud

Leaf

Leaf stalk

Old leaf

New shoot

Roots

Above: Propagation of peperomia by whole leaf and petiole cutting. **1** Cutting consists of a leaf blade and stalk. **2** Roots developing. **3** New shoot growing from cut base of leaf stalk. **4** New plant.

some of its relatives. In the notches along the leaf margins are produced plantlets, which will quickly grow when the leaf is laid flat on rooting medium.

Many succulents, such as sedum and echeveria, can be propagated by pulling off leaves and inserting them in cutting compost. A new plant will then develop from the base of each cutting.

Root cuttings

Root cuttings provide a simple and successful medium for propagating several trees, shrubs and herbaceous plants. Cuttings are taken in late winter or early spring, just before plants start making new growth. Herbaceous

Below: 1 New plants growing from wounds made in the veins of a *Begonia rex* leaf.

Leaf cuttings

plants can be lifted completely, but with trees and shrubs it is far simpler to dig down and find just one or two fairly young, strong roots. Be sure that the roots you lift are from the plant to be propagated.

Roots will vary in thickness from 3 mm (⅛ in) to 13 mm (½ in) or more, but all can be treated in the same way. Select portions of a good average diameter for the plant, and cut into pieces measuring 25–150 mm (1–6 in).

Fill a seed tray or suitable pot with John Innes seed compost to

Below: 2 Propagation of sansevieria from sectional leaf cuttings. The leaf is first cut into pieces 40–50 mm (1½–2 in) long. Cuttings are then inserted in a pot. A new plant develops from the base of the cutting.

within 25 mm (1 in) of the rim and space equally the pieces of root in rows on the surface. Cover with 13–25 mm (½–1 in) of compost, and then a piece of glass. Standing the roots in a propagator with some bottom heat will encourage rapid sprouting, but most cuttings placed in a cold frame or sheltered position outdoors will shoot equally well, if more slowly. Remove the glass cover as soon as shoots appear, and pot up or plant out in the garden when they reach 50–75 mm (2–3 in).

Hormone rooting compounds

The application of artificial root-inducing hormone, in either powder or liquid form, supplements a cutting's natural hormone content and promotes speedier root growth. Artificial hormones are applied to cuttings in very low doses—concentrations are measured in parts per million. For general garden use, an all-purpose variety will give perfectly satisfactory results and saves having to buy a range of grades. Remember that all powder preparations are prone to damp.

Dip the bottom 13 mm (½ in) of a cutting in the hormone, tap off any excess, and insert the cutting in compost. Cuttings that normally root poorly may 'take' better as a result of treatment. However, hormones have no effect on cuttings that make no attempt at all to root.

Finally, however useful, rooting hormones are no substitute for the careful selection and preparation of cutting material.

Wounding
If a thin slice of bark, 25 mm (1 in) or so long, is removed from the base of a cutting, then hormone root-inducing chemicals and water can be absorbed more readily. This is a useful trick to try with those cuttings that prove reluctant to begin growing roots, such as those of magnolia and rhododendron.

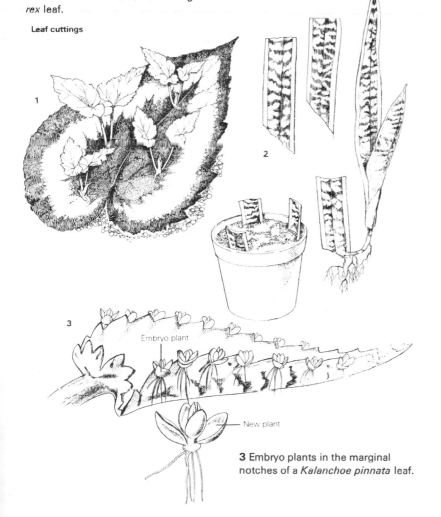

Embryo plant

New plant

3 Embryo plants in the marginal notches of a *Kalanchoe pinnata* leaf.

Above: Wounding (removing a thin slice of bark from the base of a cutting) will sometimes induce difficult subjects, such as magnolia and rhododendron, to produce roots.

Leaf bud, eye and internodal cuttings

These cuttings make thrifty use of parent plant material, but they are small and so more prone to rotting.

Leaf bud cuttings are used to propagate many evergreens, such as camellia, rhododendron and rubber plant (*Ficus elastica*). They consist of a piece of semi-ripe stem, a single bud and a leaf. Select a portion of stem with healthy, plump buds. About 25 mm (1 in) above a bud, start to make a slanting cut down under the bud, removing a shield-shaped piece of wood with bud and leaf. Do not cut into the bud. Treat the prepared cuttings with a hormone rooting mixture and insert them vertically into cutting compost, with the bud about 13 mm (½ in) below the surface. Put the cuttings into a propagator—a high bottom heat is essential for rapid rooting.

Eye cuttings. Vines are raised from eye cuttings, prepared from dormant wood just after leaf fall. Select a shoot with plump, healthy buds and cut into portions 25–50 mm (1–2 in) long, ensuring that each piece has a bud at its centre. Each piece of stem is inserted horizontally, up to half its diameter, in well-drained compost. Cuttings do not require

Above: If using John Innes seed compost to root cuttings, sprinkle a thin layer of silver sand over the surface. When a cutting is inserted, sand trickles beneath, to help aeration and drainage.

heat, but should be left in a frost-free place.

Internodal cuttings consist of a piece of stem with one leaf joint and bud (or buds), the cuts being made just above a bud and about 25 mm (1 in) below it. Softwood and semi-ripe shoots from many plants—for example, coleus and clematis—respond to this treatment. Cuttings need a well-drained compost and rooting in a heated propagator.

Preparation of containers

Containers are prepared exactly as for seed sowing (see page 27), with the exception that if John Innes seed compost is used, a thin layer of silver sand should be sprinkled over the surface. Then, when the cutting is inserted, some sand trickles into the hole

and under the base of the cutting to improve aeration. Similarly, when rooting hardwood cuttings in a clay soil outdoors, putting about 25 mm (1 in) or so of sharp sand in the base of the trench, before inserting cuttings, will improve drainage and prevent many losses from rotting.

Under mist avoid using a soil-based compost. It will waterlog and this is not conducive to root development. Rooting in pots or boxes enables established cuttings to be moved without root disturbance.

Inserting cuttings

Forcing stems of softwood and semi-ripe cuttings into compost can cause bruising. Make a small hole with a dibber, drop the cutting in and then lightly firm. Make sure that the base of the cutting is touching the compost.

Space allowed between cuttings in containers will depend on their height and spread. For instance it is quite possible to have

20 or more heather cuttings in a 76-mm (3-in) pot, while only two or three pelargonium cuttings could be inserted at most. As a general rule, try to avoid leaves touching. When they have been inserted, water cuttings using a watering can fitted with a fine rose to settle them down. No further firming should be necessary.

Rooting requirements

It is worth summarizing the main rooting requirements for various types of leafy cuttings.

1 Aim for a rooting temperature of around 21°C (70°F).

2 Keep the atmosphere as humid as possible to lessen water loss through the leaves.

3 Allow plenty of light. But under bright, sunny conditions cuttings in enclosed propagators will need shade to prevent scorching and keep the air temperature down.

4 Use a well-aerated, well-drained, disease-free compost.

5 Hygienic growing conditions are essential. Dirt breeds disease.

Care of cuttings

During rooting all cuttings should be examined daily, even those under mist. Remove dead, dying and diseased leaves and cuttings. Disease organisms find the humid, warm environment of a closed propagating case the perfect breeding ground. Surprisingly, disease is not such a serious problem under mist. The much greater light intensity, freer movement of air and the regular flushing of leaves with tepid water gives fungal spores no chance to develop.

Condensation on the inside of propagator covers should be wiped off daily. Watering, when necessary, is done with a watering can fitted with a fine rose. During warm weather most leafy cuttings benefit from being syringed with tepid water a couple of times a day. No harm is done by giving cuttings an occasional water with a general fungicide, such as benomyl, as a precautionary measure.

Buy a soil thermometer to check regularly the temperature of the rooting medium in your heated propagator. Take a reading from the level of the base of the cutting. Under no circumstances must the temperature be allowed to rise above 32°C (90°F), for excessive heat will kill cuttings as surely as cold.

The best indicator that a cutting has rooted is the appearance of new shoot growth. Once rooted, cuttings should be left in containers for no longer than absolutely necessary. Cutting composts contain little or no nutrients for new growth. Additionally, roots will grow into a solid mass, making it difficult to separate the individual cuttings.

The best time to pot cuttings is when a fibrous root system, to which the rooting medium clings, has developed. An hour or so before potting, water them well.

Keep as much soil as possible round the roots when lifting or knocking cuttings from their containers. Newly-rooted cuttings do

Once rooted, cuttings should be left no longer than necessary before potting.

Above left: Retain a good-sized ball of cutting compost round the roots.
Left: Place the rooted plant in the pot so that the 'soil line' is about 13 mm (½ in) below the rim.
Above: Lightly firm the compost. Ramming it down on to the roots will damage them.
Right: After potting, water thoroughly.

not need a rich compost at first, so use John Innes potting compost No. 1 or its soilless equivalent. Be careful not to over-pot cuttings; for almost all subjects 64-mm or 76-mm (2½-in or 3-in) pots will do. Compost should be very lightly firmed round the roots.

Any cuttings rooted one to each small pot or soil block can be grown for several weeks without potting on, provided they are fed weekly with a liquid fertilizer, such as a tomato feed.

After potting, water thoroughly. Newly-potted cuttings must be acclimatized gradually from the warm, humid environment of the propagator to the much cooler, drier conditions outside. If space allows, move newly-potted cuttings back to the propagator for a few days. Finally, before moving plants outside, place them in a cold frame for a week or two.

Plants rooted under mist must be treated with considerable care. Moving them directly to a dry environment can cause defoliation and die-back of roots, resulting in heavy losses. The ideal procedure is gradually to reduce the rate of misting, until after a week it is turned off completely. If it is not possible to 'wean' cuttings off in this way, pot them and stand them in a closed frame, maintaining a humid atmosphere. After a week transfer them to the open greenhouse.

Sometimes well-rooted cuttings will make no shoot growth, even when conditions are perfect. Such dormancy is fairly common with leaf bud cuttings of hardy evergreens. Normally, the problem can be solved by gradually hardening off the rooted cuttings and transferring them to a cold but frost-free greenhouse or frame for the winter. The following spring shoot growth usually resumes as normal.

Left: Semi-ripe cuttings of oleander root readily in water without any special treatment. The only difficulty likely to be encountered is in transferring the fleshy, brittle roots into a conventional compost without damage.

Storage of cuttings

If cuttings cannot be prepared and inserted as soon as they are collected, they can be stored. Drop the unprepared cuttings into a polythene bag, dust with a fungicide and suck out the air through a straw. Then tightly seal the bag with a twist tie and put it into the vegetable compartment of a refrigerator. At a temperature of 5°C (40°F), or thereabouts, all types of stem cuttings will keep for a month or more without deteriorating. Another way is to spray leafy cuttings with an anti-transpirant. S600, an aerosol sprayed on Christmas trees to stop the needles dropping, is suitable.

Cuttings in water

Surprisingly, a number of cuttings can be rooted by standing them in water. Stem cuttings are prepared in the normal manner. The lower third is immersed in water and the container set in a light, warm place. Water should be changed weekly. Successful subjects include: African violet from leaf cuttings; willow, forsythia and philadelphus from hardwood cuttings; box and oleander from semi-ripe wood cuttings; coleus, mint, Michaelmas daisy and impatiens from softwood cuttings.

Cuttings and their treatment

E=Eye H=Hardwood LB=Leaf Bud LC=Leaf Cuttings LS=Leaf Sections R=Root S=Softwood

S-R=Semi-ripe St=Stem St-t=Stem Tip WL=Whole Leaf

Greenhouse, bedding and houseplants from cuttings

Name Cutting types	Timing	Hormone treatment	Temp. °C (°F)
Abutilon			
S-R	Apr–Aug	Yes	21 (70)
Aechmea			
S	Apr–May	No	21 (70)
Aeonium			
WL	May–Jul	No	15 (60)
Root on open bench			
Aphelandra			
S, S-R	Jun–Aug	Yes	21 (70)
Begonia			
S	Apr–Jul	Yes	21 (70)
Begonia rex			
WL, LS	Apr–Jul	Yes	21 (70)
Beloperone (Shrimp plant)			
S-R	Jun–Aug	Yes	15 (60)
Also roots well from stem tips about 50 mm (2 in) long			
Cacti			
	May–Jul	No	21 (70)
All shoots very fleshy Dry base of cutting in air. Rest on compost surface. Do not insert			
Camellia			
LB, S-R	Aug	Yes	21 (70)
Leaf bud cuttings from semi-ripe shoots			
Carnation			
S	Dec–Feb	Yes	15 (60)
Chrysanthemum			
S	Feb–Jun	Yes	15 (60)
Time depends on variety and flowering time			
Cissus (Kangaroo vine)			
S	Jun–Aug	Yes	15 (60)
Citrus (Lemon, Lime, etc.)			
S-R	Jul–Aug	Yes	21 (70)
Take with heel if possible			

Name	Timing	Hormone	Temp.
Codiaeum (Croton)			
S-R	Jun–Aug	Yes	24 (75)
Coleus			
S	All year	Yes	15 (60)
Dahlia			
S	Mar–May	Yes	15 (60)
Dianthus, see Carnation			
Dieffenbachia (Dumb cane)			
S-R	All year	Yes	21 (70)
Echeveria			
WL	May–Jul	No	15 (60)
Root on open bench			
Epiphyllum (Christmas cactus)			
St	May–Jul	No	15 (60)
Break off portions of stems at joints			
Euphorbia pulcherrima (Poinsettia)			
S-R	Apr–May	Yes	21 (70)
Euphorbia splendens (Crown of thorns)			
St-t	May–Jul	Yes	15 (60)
Dry base of cutting before inserting			
Fathedera			
S-R	Jul–Aug	Yes	15 (60)
Ficus (Rubber plant)			
S-R, LB	May–Jul	Yes	21 (70)
Large-leaved. Dry off sap before inserting (see page 50)			
Fuchsia			
S	All year	Yes	15 (60)
Gardenia			
S-R	Jun–Jul	Yes	24 (75)
Wound base of cutting			
Hedera (Ivy)			
S, S-R	May–Aug	Yes	15 (60)
Heliotropium			
S	May–Jul	Yes	15 (60)
Hydrangea			
S	May–Jun	No	15 (60)
Impatiens (Busy Lizzie)			
S	Apr–Sep	Yes	15 (60)
Kalanchoe			
S	May–Jul	Yes	15 (60)

Name	Timing	Hormone	Temp.
Some form small plantlets on their leaves			
Lantana			
S	May–Jul	Yes	15 (60)
Maranta (Prayer plant)			
St-t	Apr–Sep	No	21 (70)
Monstera (Swiss cheese plant)			
St	All year	Yes	21 (70)
Pelargonium ('Geranium')			
S, S-R	Feb–Oct	Yes	15 (60)
Do not dry base before inserting			
Peperomia (Pepper plant)			
S	May–Jul	Yes	21 (70)
Whole leaf with stalk attached. New plant grows from cut surface of leaf stalk			
Philodendron			
St-t	All year	Yes	24 (75)
Pilea			
S	All year	Yes	21 (70)
Plumbago			
S-R	Jun–Aug	Yes	21 (70)
Take with heel			
Saintpaulia (African violet)			
	May–Jul	No	15 (60)
Whole leaf with stalk attached. New plant grows from cut surface of leaf stalk. Roots well in water			
Sansevieria (Mother-in-law's tongue)			
LC	May–Jul	No	21 (70)
See page 40–41 Variety 'Laurentii' loses yellow leaf margins			
Streptocarpus			
LS	Jun–Jul	No	21 (70)
See page 40. New plant grows from cut surface			
Tradescantia (Wandering Jew)			
St	All year	No	15 (60)
Tropaeolum			
S	Jun–Jul	No	24 (75)
Zebrina, see Tradescantia			

Hardy herbaceous plants from cuttings

Name Cutting types	Timing	Hormone treatment	Temp. °C (°F)
Acanthus			
R	Oct–Mar	No	No heat
Needs frost protection			
Achillea (Yarrow)			
S	May–Jul	Yes	15 (60)
Anchusa			
S	Apr–May	Yes	15 (60)
R	Oct–Mar	No	No heat
Needs frost protection			
Anemone			
R	Oct–Mar	No	No heat
Needs frost protection			
Anthemis (Camomile)			
S	May–Jul	Yes	15 (60)
Armeria (Thrift)			
S	Jun–Jul	Yes	15 (60)
Artemisia (Southernwood)			
S	Jun–Jul	Yes	15 (60)
Take with heel			
Aster			
S	Apr–May	Yes	No heat
Needs frost protection. Make from basal shoots			
Campanula			
S	Jun–Jul	Yes	15 (60)
Chrysanthemum			
S	May–Jul	Yes	15 (60)
Coreopsis			
S	Jul–Aug	Yes	15 (60)
Take with heel if possible			
Delphinium			
S	Apr–May	No	No heat
Needs frost protection. Make from basal shoots			
Dianthus			
S	Jun–Aug	Yes	15 (60)
Take with heel			
Dicentra (Bleeding heart)			
S	May–Jun	Yes	15 (60)
R	May–Jun	No	No heat
Gypsophila			
S	Apr–Jun	Yes	15 (60)
Helenium			
S	Jun–Jul	Yes	15 (60)
Helichrysum (Everlasting flower)			
S-R	Jul–Aug	No	No heat
Needs frost protection until rooted			
Lamium (Deadnettle)			
S	Apr–Sep	No	No heat
Needs frost protection			
Limonium			
R	Oct–Mar	No	No heat
Needs frost protection			

Name	Timing	Hormone	Temp. °C (°F)
Linum			
S	Jun–Aug	No	No heat
Needs frost protection			
Lobelia			
S	May–Jun	Yes	15 (60)
Lupinus (Lupin)			
S	Apr–Jun	No	No heat
Make from basal shoots			
Lysimachia (Creeping Jenny)			
S	Mar–May	No	15 (60)
Lythrum (Purple Loosestrife)			
S-R	Jul–Aug	No	No heat
Mentha (Mint)			
S	Apr–Sep	No	No heat
Mimulus (Musk)			
S	Apr–Sep	Yes	15 (60)
Nepeta (Catmint)			
S-R	Jul–Aug	No	No heat
Oenothera (Evening primrose)			
S	Jun–Aug	Yes	15 (60)
Paeonia (Peony)			
S	Mar–May	No	No heat
Make from basal shoots			
Papaver orientale			
R	Oct–Mar	No	No heat
Needs frost protection			
Penstemon			
S	Jun–Aug	No	No heat
Phlox			
R	Sep–Nov	No	No heat
Needs frost protection			
S	Apr–Jun	No	No heat
Make from basal shoots			
Primula denticulata			
R	Oct–Feb	No	No heat
Needs frost protection			
Pyrethrum, see Chrysanthemum			
Scabiosa			
S	Apr–Jun	No	No heat
Make from basal shoots			
Sedum			
S	Jul–Aug	Yes	15 (60)
Stokesia			
S	Mar–May	No	15 (60)
Symphytum (Comfrey)			
R	Oct–Mar	No	No heat
Needs frost protection			
Verbascum			
R	Oct–Mar	No	No heat
Needs frost protection			
Veronica			
S	Apr–Jun	Yes	15 (60)
Viola			
S	May–Aug	No	15 (60)

Trees and shrubs from cuttings

Name Cutting types	Timing	Hormone treatment	Temp. °C (°F)
Abelia			
SR	Jul–Aug	Yes	21 (70)
	Nov–Dec	No	No heat
Root in frame			
Abutilon			
S	Apr–Aug	Yes	21 (70)
Azalea, see Rhododendron			
Berberis			
S, S-R	Jun–Aug	Yes	21 (70)
Take with heel			
H	Nov–Dec	Yes	No heat
Take with heel. Root in frame			
Buddleia			
S	May–Jul	No	15 (60)
Buxus (Box)			
S	Jun–Aug	Yes	15 (60)
S-R	Aug–Sep	Yes	No heat
Take with heel. Root in frame			
Callistemon (Bottle brush)			
S-R	Aug	Yes	15 (60)
Ceanothus			
S	Jun	Yes	15 (60)
Prefers mist			
H	Nov–Dec	Yes	No heat
Root in frame			
Chaenomeles (Quince)			
H	Oct–Nov	Yes	15 (60)
Root in propagator			
Choisya			
S-R	Jul–Aug	Yes	15 (60)
Cistus			
S, S-R	Jul–Aug	Yes	15 (60)
Clematis			
S-R	Jun–Jul	No	15 (60)
Conifers, see page 38			
Cornus (Dogwood)			
S-R	Jul–Aug	No	15 (60)
H	Oct–Dec	No	No heat
Root in open ground			
Cotoneaster			
S-R	Jul–Aug	Yes	15 (60)
Responds well to mist			
Cytisus (Broom)			
S-R	Jul	Yes	21 (70)
Prefers mist and bottom heat			
Daphne			
S-R	Jul	Yes	21 (70)
Prefers mist and peat/sand compost			
Elaeagnus			
S-R	Jul–Aug	Yes	15 (60)
Escallonia			
S-R	Jul–Aug	Yes	15 (60)
H	Nov–Dec	No	No heat
Root in open ground			

Euonymus (Spindle bush)
H   Nov–Dec   No   No heat
Deciduous. Root in frame
S-R   Jul–Aug   No   15 (60)
Evergreen

Forsythia
S   May–Jun   Yes   21 (70)
H   Nov–Dec   No   No heat
Root in open ground

Fuchsia
S   All year   Yes   15 (60)
Take cuttings from hardy types May–Jul

Gaultheria
S-R   Jul–Aug   Yes   15 (60)

Genista
S-R   Jul–Aug   Yes   15 (60)
Take with heel. Will root in unheated frame

Griselinia
S-R   Jul   Yes   15 (60)
Prefers mist
H   Oct–Nov   Yes   No heat
Root in frame

Heather
S-R   Sep–Oct   No   No heat
Root in frame

Hebe
S-R   Jul–Aug   No   15 (60)
Will root in unheated frame

Hibiscus rosa-sinensis
S-R   Jul–Aug   Yes   15 (60)
H   Oct–Nov   Yes   15 (60)
Bottom heat only

Hibiscus syriacus
S   Jul   Yes   15 (60)

Hydrangea
S   May–Jun   No   15 (60)

Hypericum (St John's Wort)
S   Jun–Jul   No   15 (60)

Ilex (Holly)
S-R   Aug–Sep   Yes   24 (75)
Prefers mist

Kalmia
S-R   Aug   Yes   15 (60)
Will root in unheated frame

Kerria
H   Nov–Dec   No   No heat
Root in open ground

Lavandula (Lavender)
S-R   Aug   No   No heat
Root in frame

Ligustrum (Privet)
H   Oct–Nov   No   No heat
Root in open ground

Lonicera (Honeysuckle)
H   Apr   No   No heat
Root in open ground

Magnolia
S-R   Jul   Yes   24 (75)
Take cuttings from young plants if possible. Wound cuttings at base

Mahonia
S-R   Jul–Aug   Yes   21 (70)
Prefers mist

Oleander
S-R   Aug   Yes   21 (70)

Olearia
S-R   Aug–Sep   Yes   No heat
Root in frame

Parthenocissus (Virginia creeper)
S   Aug   Yes   21 (70)
H   Apr   No   No heat
Root in frame

Philadelphus (Mock orange)
H   Mar   No   No heat
Root in open ground

Pieris
S-R   Jul–Aug   Yes   21 (70)

Potentilla
S-R   Jul–Aug   No   No heat
Take with heel

Prunus cerasifera
H   Nov–Dec   No   No heat
Root in open ground

Prunus laurocerasus
S-R   Jul–Aug   Yes   15 (60)

Prunus lusitanica
S-R   Jul–Aug   Yes   15 (60)

Pyracantha
S-R   Jul–Aug   Yes   21 (70)
Take with heel

Rhododendron
S-R   Jul   Yes   24 (75)
Wound cuttings. Root in mix of equal parts perlite and peat

Ribes (Flowering currant)
S   Jun–Jul   Yes   15 (60)

Rosa (Rose), see page 39

Rosmarinus (Rosemary)
S-R   Jul–Aug   Yes   24 (75)
Prepare from semi-ripe wood

Salvia (Sage)
S   Jun–Aug   Yes   15 (60)

Senecio
S-R   Jul–Aug   Yes   15 (60)
Will root slowly in cold frame

Skimmia
S-R   Jul–Aug   Yes   15 (60)

Symphoricarpos (Snowberry)
H   Oct   No   No heat
Root in open ground

Syringa (Lilac)
S   May   Yes   21 (70)
Rarely successful unless rooted under mist
S-R   Jun–Jul   Yes   15 (60)
Take with heel

Tamarix (Tamarisk)
H   Mar   No   No heat

Thymus (Thyme)
S   Jun–Jul   No   No heat
Take with heel. Root in frame

Viburnum
S   May–Jun   Yes   21 (70)
Must have good roots before potting

Vitis (Vine)
E   Feb   No   15 (60)
See page 42
S-R   Jul–Aug   Yes   15 (60)

Weigela
S-R   Jun–Jul   Yes   15 (60)
Take with heel

Wisteria
S-R   Jul–Aug   Yes   21 (70)
H   Mar   No   15 (60)

Fruit from cuttings

Name Cutting types	Timing	Hormone treatment	Temp. °C (°F)

Citrus (Lemon, Lime, etc.)
S-R   Jul–Aug   Yes   21 (70)
Take with heel if possible

Currant, Black (Ribes sp)
H   Oct–Nov   No   Root outdoors
Use current year's growth. Make 200 mm (8 in) long. Insert 150 mm (6 in) deep

Currant, Red
H   Oct–Nov   No   Root outdoors
As above. Make cuttings 300 mm (12 in). Remove all buds but top four

Currant, White
H   Oct–Nov   No   Root outdoors
As for Red Currant

Fig (Ficus carica)
H   Oct–Nov   No   10 (50)
Store cuttings in moist sand over winter. Insert Mar–Apr

Gooseberry (Ribes grossularia)
H   Oct–Nov   No   Root outdoors
See Red Currant

Grape (Vitus vinifera)
H   Oct–Nov   No   Root outdoors
See Fig
S   Mar–May   Yes   24 (75)
Root in 10–12 days under mist

Quince (Cydonia oblonga)
H   Nov–Dec   No   Root outdoors
Use one-year-old wood with heel, or prepare from two- to three-year-old wood. See Fig

Layering

A layer is a stem induced to produce roots while it is still attached to the parent plant. Therefore, unlike a cutting, a layer has the advantage of being continually supplied with nutrients and water until it has developed enough roots to be separated and grown on its own. Thus many plants difficult to raise from cuttings can be propagated by layering. Also, as layered shoots tend to be bigger than cuttings, it is possible to raise larger plants in a shorter time.

Root formation

Root development on a layer, as on a cutting, depends on a moderate temperature, continuous moisture (the rooting medium must not be allowed to dry out) and good aeration.

Development of roots is stimulated by checking the flow of sap so that it collects at the point where roots are to form. Sap flow can be arrested by bending, twisting, notching, girdling or applying a wire tourniquet to the stem.

Hormone rooting powders help to stimulate root development, and are particularly worthwhile when layering difficult subjects, magnolias for instance.

When the newly-rooted cutting has been separated from the parent and potted, it needs to be kept in a closed case or frame for a few weeks until it has formed a strong, vigorous root system. The larger a layer, the more difficult it is to establish after separation, so as a rule keep layers shorter than 200 mm (8 in).

Timing

Late spring to early summer is the best time to layer indoor and outdoor plants. At this time of year growth is very active and root formation on healthy, strong

plants should be fairly rapid. Cold inhibits root growth, so autumn or winter layers rarely start to root before spring and consequently often rot.

Techniques

There are a multitude of layering methods, but for raising just a few plants the simplest and most successful are:

    Simple layering
    Tip layering
    Serpentine layering
    Air layering

Simple layering is accomplished by bending a shoot down to ground level, covering the bend with soil, leaving the tip exposed, and allowing roots to develop.

Outdoors, the operation can be carried out in the early spring, selecting a flexible, one-year-old growth. Or it can be done a little later with a shoot from the current season's growth, choosing one that is semi-ripe (pliable but not soft and sappy).

Garden soil will need to be well

drained but able to hold moisture. If it contains clay, work in a mixture of peat and sharp sand to a depth of 150 mm (6 in): if the tendency is to sandy conditions, fork in a liberal dressing of peat to a similar depth to aid water retention. A handful of bonemeal worked into the top 50 mm (2 in) will stimulate root growth once this has begun.

The shoot must be wounded in the region where it is to be bent. Either make a long slanting cut, penetrating about half-way, or remove a ring of rind 13 mm (½ in) wide. Make a shallow depression in the soil, bend the layer gently down into it, securely fix with a long wooden peg or wire staple, and cover to a depth of about 25 mm (1 in). Support the shoot tip with a small stake and tie.

Rooting time varies, but as a rule most layers made in spring can be separated, potted and moved to a cold frame or unheated greenhouse in late autumn. A few subjects, such as magnolias and rhododendrons, are best left

Top: Serpentine layering. A single long, pliable shoot is pegged down at regular intervals, either directly into the soil or into pots with John Innes potting compost No. 1. When roots have formed, the layered stem can be lifted out and separated into rooted portions, each with a bud or developing shoot. These can then be planted out

Left: Simple layering. Four steps are shown. First, a cut is made in the shoot in the region where roots are to form. The cut layer is then carefully bent down into a shallow trench. It is held securely in the ground with the aid of a long peg, and the shoot tip is supported by a stake. When well rooted the layer is then separated from the parent plant. Root formation can take from 6 to 24 months.

Above: Air layering a rubber plant. Select a healthy shoot and remove one or two leaves approximately 300 mm (12 in) from the tip. From just under a leaf joint, make a long slanting cut. Keep the two surfaces apart with a slither of wood.

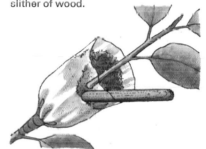

Above: Dust the wound with a hormone rooting powder. A ball of moist sphagnum moss is wrapped around the wound and then enclosed within a polythene sleeve.

for another season (i.e. 18 months in all) before being parted from the parent plants, as rooting is slow.

Greenhouse plants such as the dumb cane (dieffenbachia) develop long, leggy stems ideal for simple layering. Fill a pot with cutting or seed compost, bend the stem down and peg firmly in position. Pot rooting is almost always successful under glass. Layers can be rooted outdoors in pots, provided that watering is not neglected.

Tip layering is similar to simple layering except that rooting takes place near the tip of the current season's shoot growth. It is a natural method of propagation for many trailing fruits, such as blackberries and loganberries.

Layers are made from mid-summer when canes just begin to arch. Soil is prepared as for simple layering, and a depression 50–75 mm (2–3 in) deep is made. The shoot tip is placed into this, covered with soil and lightly firmed. The buried tip readily roots and produces a vigorous young

shoot. By late October the rooted layer is ready to be separated. Lift with care to avoid damaging the tender new shoot. In the case of a cane, the remainder of the shoot is either cut back hard or tied to a support and fruited the following season. Pot the severed layer and stand it in a cold frame or plant out in a well sheltered position.

Serpentine layering. A long single shoot, of clematis for instance, is layered every 450–600 mm (18–24 in) according to its pliability and the positions of its buds. Each exposed portion of stem must have at least one bud to grow into a new shoot. Layers can be pegged into pots or directly into the soil. As stems selected for serpentine layering must be thin, wounding is carried out by making a slanting cut immediately behind a node, and to just about half-way through the shoot, or by twisting, which tends to split the rind. Layers, normally well rooted by autumn, are lifted with care and separated into portions, each with a bud or shoot.

Above: The sleeve is sealed at both ends with waterproof tape to prevent moisture being lost from the moss. The layer is separated when roots can be clearly seen through the polythene.

49

HBP

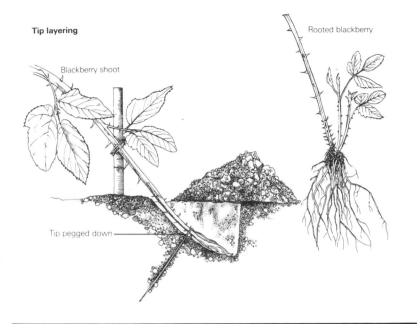

Tip layering

Blackberry shoot

Rooted blackberry

Tip pegged down

Left: Propagating a blackberry by tip layering. Peg down the tip of a long new cane into a shallow depression ready for covering with firmed soil. The rooted shoot is separated from its parent (right) and transplanted. An old piece of cane is left for easy handling.

Air layering is the process by which roots are induced to grow on an aerial stem. It is a technique normally adopted for propagating leggy indoor trees and shrubs, but can be applied to those grown outdoors. For the best chance of success, select a previous season's growth for layering, and carry out the operation in the spring. On a rubber plant, select a healthy shoot 13–20 mm (½–¾ in) thick. At a point approximately 300 mm (12 in) from the tip, and just under a node where the leaf has been removed, make a long slanting upward cut half-way through the stem, or remove a ring of bark about 13 mm (½ in) wide. Keep the two surfaces of the slanting cut apart by inserting a slither of wood, forming a bird's-mouth. The wound will exude a thick, sticky white sap which will congeal and can be scraped off before liberally applying hormone rooting powder. A ball of moist—but not saturated—sphagnum moss about 50 mm (2 in) thick is wrapped around the wound and then enclosed within a polythene sleeve. The sleeve is sealed at both ends with a waterproof tape, raffia or fillis twine. This will prevent moisture being lost from the moss and water from entering. Support the layered shoot by staking or tying it to a neighbouring stem. Roots take a couple of months to develop and the layer is ready to pot when they can be clearly observed through the polythene wrap. When a layer has taken a long time to root, shoot pruning may be necessary to keep leaf and stem growth in proportion to that of the roots. Outdoors, rooted air layers are separated from the parent plant while dormant.

Plant	Time of year	Type of layer	Rooting time	Comments
Blackcurrant/ Loganberry	Jul–Aug	Tip	12–16 weeks	Separate Nov–Dec
Carnation (border)	Jul–Aug	Simple	8–10 weeks	Pot rooted layers and stand in frame
Clematis	Apr–May	Serpentine	20–24 weeks	Select two-year-old stems
Holly	Jun	Air	12–16 weeks	Plants up to 600 mm (2 ft) produced in first season
Lapageria rosea	Apr–May	Serpentine	8–12 weeks	Cool greenhouse climber. Pot rooted layers and stand in propagator
Lilac	Apr–May	Simple	20–24 weeks	One- or two-year-old shoots can be used
Magnolia	Apr–May	Simple or air	18–24 months	Use one- or two-year-old shoots. Root development is very slow
Rhododendron/ Azalea	Apr–May	Simple or air	6–24 months	Rooting rate depends on type
Rubber plant	Apr–May	Air	12–16 weeks	
Viburnum	Apr–May or Jul–Aug	Simple or air	18–24 months	Use one-year-old wood (Apr–May); semi-ripe wood (Jul–Aug)
Vine (including grape)	Apr–May	Simple or serpentine	20–24 weeks	Use one- or two-year-old wood

Bulbs and Co.

Bulbs, corms, tubers, rhizomes and pseudobulbs are all specialized plant organs which serve the same purpose. They act as a food store, enabling plants to survive during a dormant period when growing conditions are far from ideal, and a means of propagation.

Bulbs

A bulb is a type of shoot composed of fleshy leaves closely packed together on a very flattened stem called the basal plate. There are two bulb structures:

1 *Tunicate*, in which the bulb is made up of continuous fleshy leaves protected by a dry outer skin — the tunic. Examples are the tulip, daffodil and hyacinth.

2 *Non-tunicate*, in which the bulb is composed of many separate scale-shaped leaves, each attached to the basal plate but not protected by a tunic — e.g. the lily.

Bulbs can be increased by means of: offsets; bulbils and bulblets; basal cuttage; scaling; and bulb cuttings.

Offsets. Many bulbs continually increase themselves by means of buds, situated on the basal plate between the fleshy leaves. These develop into new bulbs called offsets which are removed when the bulbs are lifted. At planting time they can be placed either in pots or trays, or in the open ground. Large, plump offsets will often reach flowering size within a year, whereas small, flat ones may take three or four.

Bulbils and bulblets. Many lilies—e.g. *Lilium sulphureum*—produce small bulbs called bulbils in the axils of their leaves. These can be removed and planted in a tray filled with John Innes potting compost No. 1. When placed in a cool greenhouse or frame, they will rapidly increase in size. It takes three to four years to produce flowering sized bulbs. Small bulbs that develop on the flowering stem below ground are called bulblets, and are also characteristic of many lilies—e.g. *L. umbellatum, L. hollandium and L. longiflorum.*

Basal cuttage. Hyacinths do not readily produce offsets, but can be induced to form bulblets by 'scooping' or 'scoring'.

Bulbs are lifted for treatment immediately after the foliage has died down. 'Scooping' involves removing the entire basal plate to expose the fleshy storage leaves. 'Scoring' exposes the bulb's leaves by three or four cuts made right through the basal plate. Otherwise a bulb can be scooped and

Above: Many bulbs continually produce lateral bulbs, called offsets, which can be split off. Depending on size, offsets take from one to four years to flower.

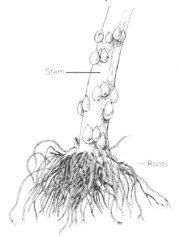

Stem

Roots

Above: Small bulbs produced by some lilies on the flowering stem below ground are called bulblets.

Leaves

Bulbils in leaf axils

Above: Small bulbs in leaf axils are called bulbils. Removed and planted in John Innes potting compost No. 1, these reach flowering size in two to four years.

Common plants with specialized stems and roots

Bulbs	Corms	Tubers	Rhizomes
Allium	Colchicum	Agapanthus	Achemenes
Amaryllis	Crocus	Anemone	Canna
Chionodoxa	Freesia	Arum	Convollaria
Clivia	Gladiolus	Begonia	Iris
Crinum	Ixia	Caladium	Monbretia
Erythronium	Strebergia	Claytonia	Trillium
Eucharis	Tritonia	Cyclamen	Zantedeschia
Fritillaria		Dahlia	Zingiber
Galanthus		Dicentra	
Haemanthus		Eranthus	
Hippeastrum		Eremurus	
Hyacinthus		Gloriosa	
Hymenocallis		Gloxinia	
Iris		Ranunculus	
Leucojum		Tropaeolum	
Lilium	Tigridia		
Musari	Tulipa		
Narcissus	Vallota		
Nerine	Veltheimia		
Oxalis	Watsonia		
Scilla	Zephyranthes		

then scored. Dust the cuts with a fungicide and stand the bulbs, base down, on a tray filled with dry sharp sand. Leave them for a week or two, and then slightly moisten the sand. Keep the temperature level to at least 21°C (70°F). After two or three months bulblets will have formed on the exposed leaves or score lines. Outdoors, plant the bulb upside-down in the autumn, covering it with 75–100 mm (3–4 in) of soil.

Below left: Basal cuttage. First the entire basal plate is removed by scooping (top). The basal plate is then scored (middle) and bulblets will form on score lines (bottom).

Below: Scaling. Scales are removed from a lily bulb (top) and inserted in a tray of prepared compost (middle). Rooted scale with bulblet at base (bottom).

Below right: Mature corm with cormels (top). Separating cormels from the parent corm (bottom).

Bulb cuttings. Many bulbs— e.g. nerine, scilla, hippeastrum and sperkelia—can be increased by cutting them into segments, each with a piece of basal plate, and planting them vertically in cutting compost. Bulblets will develop along the basal plate. Cuttings are prepared at the start of the resting period, and treated in the same way as lily scales.

The mother bulb will slowly rot away, and the following autumn the small bulbs can be planted in rows. Bulbs are then lifted annually and replanted until they reach flowering size, 125–150 mm (5–6 in) in diameter.

Scaling. Scales make a type of leaf cutting and provide an easy method of increasing lilies. Lift mature bulbs after flowering and carefully pick off the outer two rows of scales only. Replant the mother bulb. Insert the scales upright in a tray filled with cutting compost, 40 mm (1½ in) apart.

Small bulblets and roots will form on the base of the scales after six to eight weeks. For rapid rooting keep the temperature at 18°C (65°F). Rooted scales can be lined out in open ground the following spring, and will reach flowering size in four to five years. Garlic bulb scales (cloves) can be planted straight in the open.

Corms

A corm is a solid swollen stem covered by dry scale-like leaves— the tunic—with an apical shoot and a number of buds. Examples are gladiolus and crocus. Propagation is by means of new corms, cormels or division. New corms of flowering size simply develop on top of the corm grown during the previous season. After lifting and drying, the new corms are separated from the old ones and stored.

Cormels are miniature corms

that develop round the junction between the old and new corm. These are separated from the parent and stored in moist peat through the winter. The following season they are planted in the open 100 mm (4 in) apart in a drill 50 mm (2 in) deep. Cormels will then produce a larger corm, but one more year of growth is needed to attain flowering size.

Below left: Tuberous begonias can be increased by division (top), but each portion must have at least one bud. Root tubers of dahlia are divided in the same way (bottom).

Below: Flag iris is increased by cutting off portions of rhizome about 150 mm (6 in) long, ensuring that each piece has a shoot. Leaves should be trimmed back.

Right: Removing pseudobulbs from a cymbidium orchid immediately after it has finished flowering.

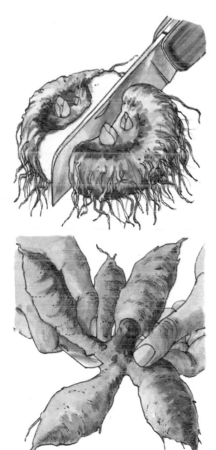

Large corms can be divided into segments, provided that each piece has a bud. All cut surfaces should be dusted with fungicide.

Tubers

A tuber can be a swollen stem at ground level (e.g. begonia, cyclamen and gloxinia); or below ground level (e.g. potato, caladium and Jerusalem artichoke). Such stems have buds ('eyes') which can develop into new shoots. It can also be a swollen root with buds only on the crown (e.g. dahlia).

Propagation of tubers is by division or, as in the case of dahlias, by cuttings, removing new shoots when they are 75–100 mm (3–4 in) long.

Stem tubers must be divided so that each portion contains at least one bud. Root tubers must have a stem or segment of crown on each piece. A few plants—e.g. *Begonia sutherlandii*—produce small aerial tubers called tubercles in the axils of their leaves. These can be grown in the same way as full-sized tubers.

Rhizomes

Rhizomes are specialized stems which grow horizontally at or below the soil surface. They can be thick, fleshy and form clumps (e.g. flag iris and canna). Other types are thin, slender and spreading (e.g. lily-of-the-valley), or tuberous-like in structure (e.g. montbretia and achimenes).

Propagation of all types is by simple division undertaken at the end of the growing season, in early spring or immediately after flowering. With thick, fleshy rhizomes, portions about 150 mm (6 in) long are cut off, ensuring that each piece has a bud or developing shoot. Any well formed leaves can be cut back. Thin rhizomes are handled in much the same way. Portions of rhizome with at least one large bud (called a 'pip') are removed for replanting. The central portion of a

53

large clump, consisting mainly of old growth, is normally discarded. The tuber-like rhizomes of achimenes can be divided into almost seed-sized portions. In some species structures virtually identical to the rhizomes develop in the leaf axils and may also be used for propagation.

Pseudobulbs

Pseudobulbs are a characteristic of many common orchids. They are swollen pieces of stem which are produced at or above soil level, and can vary from the thickness of a thin pencil to a fat bulb. When repotting orchids im-

mediately after flowering, or at the start of their growing season, pseudobulbs can be cut off and potted up in a suitable compost. If they have not already developed, a shoot will grow from either the base or top end of the pseudobulb, and roots from the base of this shoot. In the case of cymbidium, which has a very succulent pseudobulb, a well-rooted shoot can be removed from the bulb and the bulb induced to develop a second shoot. When the pseudobulb is old and leafless, it is called a back bulb, and a green bulb is a pseudobulb with leaves. Both can occur on the same plant and be used for propagation.

Below: **1** Tomatoes are grafted when both the rootstock and scion (fruiting) plants are about 150 mm (6 in) high. Remove the seed leaves from both.
2 Remove the top of the rootstock and make a downward cut into the stump. A corresponding upward cut is made into the stem of the scion.
3 Carefully join the two plants together by interlocking the two tongues.
4 Firmly hold the graft in position, binding it with sticky tape.
5 Pot the two plants into a single container, and support with a stake and tie.
6 When the graft has taken and healed, normally after a period of about 24 days, cut away the roots of the scion variety.

Grafting and budding

We shall look only briefly at propagation by means of grafting and budding, examining just two subjects in detail—cacti and tomatoes. The simple reason for this is the general lack of suitable rootstocks available in small numbers to amateur gardeners. But as many plants bought for gardens are either grafted or budded, it is worth looking at the reasons for these propagation techniques.

Grafting

Grafting is the art of uniting two portions from different plants so that they grow together as one. The top part, which develops into the new plant, is called the scion and the bottom, which forms the root system, is called the stock. Budding is a variation of this in which a single bud, still called the scion, is grafted on to a rootstock; the bud then develops into the new plant.

The most important reasons for grafting and budding are the following.

1 To obtain the advantages of certain rootstocks. Many plants grown on their own roots are unable to tolerate certain growing conditions, or are prone to soilborne problems. For example, many tomato varieties are attacked through their roots by destructive soil-borne organisms. Grafting them on to a rootstock resistant to these pests and diseases is a simple method of overcoming the problem. A rootstock can strongly influence the growth of a plant grafted on to it. Apple varieties, for instance, are grafted on to carefully-selected rootstocks so that growth may be controlled—e.g. Malling XI rootstock gives rise to quite a small tree while Malling 106 has only a semi-dwarfing influence. Hybrid tea and floribunda roses are budded on to rootstocks in order to ensure some resistance to certain soil pests and cold damage in winter.

2 To obtain special forms of growth. Many types of weeping cherry, standard roses and family

Above: Some variegated plants are grafted on to all-green close relatives to hasten growth, which would be painfully slow if they had to rely on their own roots.

fruit trees (which have several varieties of either apples or pears growing from a single tree) are all grafted or budded.

3 To increase plants that cannot be easily raised by other means of propagation. Many varieties of plants—lilac, blue spruce, magnolia, walnut and maple, to name just a few—will not root easily as cuttings. Grafting is therefore the only practical means of increasing them in large numbers. An easily-raised close relation is normally used as the rootstock—e.g. the rootstock for French lilacs is the common lilac, *Syringus vulgaris.*

4 To hasten growth. Some plants, many cacti for example, grow very slowly on their own roots. By using a more vigorous close relation as a rootstock, it is possible to hasten growth and shorten the period to flowering and fruiting.

Less common applications of grafting and budding are:

● to change the variety of an established fruit tree. This is cut back very hard, leaving just the trunk and a few sturdy branches. A new variety is then grafted on.

● to aid pollination. For ex-

ample, holly has separate male and female plants and pollination must be carried out before berries form on the female. If you have only a female bush the pollination problem can be overcome by grafting one or two shoots from a male bush on to it.

● to repair damaged trees. When a tree's trunk has been damaged but its roots are sound, it is sometimes possible to effect a repair by bridging the damaged region with new shoots grafted in.

For a graft to be successful, the stock and scion must first of all be compatible. Next, the two cut surfaces to be united must be brought into close contact so that their cambial regions (areas where there is very active cell division) are touching. If the union does not heal, the scion will be unable to obtain nutrients and water via the stock, and will die. To ensure against this, the stock and scion must be held firmly during the healing process. The optimum time for grafting is normally spring, when growth is most active. For success, the temperature should average 13°C (55°F) at least.

Grafting tomatoes

Putting your favourite tomatoes on to new roots could solve many problems of disease that inflict plants, for numerous tomato varieties are vulnerable to attack via the roots. Suitable disease-resistant rootstocks can be raised from seed, which is available through most seedsmen. By far the best rootstock variety goes by the initials 'KNVF'. It is resistant to damage by corky root, root knot eelworm and the most destructive wilt diseases. Rootstock varieties are slower to germinate than fruiting ones, and must be started a week earlier. Seed is sown thinly, germinated at 18°C (65°F) and seedlings, when large enough, are pricked out one to each small pot.

Plants are grafted when they reach about 150 mm (6 in) high. The graft used is called the 'tongue approach'. Using a new razor blade or very sharp knife, remove the seed leaves from the rootstock plant and the fruiting plant. Make a downward cut in the rootstock plant and a corresponding but upward cut in the fruiting plant. Each cut should penetrate about half-way through the stem. The two plants are carefully joined together by interlocking the two tongues, and firmly held in position by binding with sticky tape. The two grafted plants are potted into a single pot, and placed in a propagator where a temperature of around 23°C (75°F) can be maintained. Union takes 12–14 days, but allow another 10 days before removing the tape. Finally, cut away the non-resistant roots of the scion (fruiting) variety and the top of the rootstock variety.

A modified and slightly easier method of grafting is to remove the top of the rootstock variety, and make the downward cut into the stump. The rest of the procedure remains as described.

After making each graft, sterilize the knife or razor blade used by dipping it in methylated spirits or passing it through a flame. The graft union, even when healed, is a weak point, and plants will need firm support throughout their lives.

Cacti

Unlike tomatoes, grafting cacti is more for fun than profit, although commercial growers do use the technique for rapidly raising slow-growing species. Another advantage of grafting is to keep alive special forms and abnormalities such as albinos, the red cap cacti *(Gymnocalcium mihanovichii)* and the crested varieties that rarely survive in nature. These often contain little or no chlorophyll, so they cannot manufacture food. Left to their own devices these would die. But grafted on a normal, healthy green plant, they happily lead a parasitic type of existence.

These oddities of the cactus world are expensive to buy, but if you can beg one or two off-shoots, it is very easy to graft your own plants. The most common cactus used as a rootstock is the triangular *Hylocerus guatemalens,* but it dislikes temperatures much below 10°C (50°F), and rotting then becomes a problem. Both *Econopsis* and *Trichocerus* are far more tolerant of cold.

The grafting technique is simple. Take the stock plant—it should be about 150 mm (6 in) high—and slice off the top 25 mm (1 in) or so with a sharp knife. Next, bevel the edge of the stock, trimming back no further than the outside ring of vascular tissue which is clearly visible. Take a piece of cactus to be grafted on to the stock, and cut a thin, clean, horizontal slice off its base to expose the vascular tissue. Bevel the edges. Carefully place the

scion on the stock, ensuring that the two rings of vascular tissue touch in one place at least. If they fail to touch, connecting tissue will not develop. Hold the scion firmly on to the stock and stretch a couple of elastic bands over the top of the scion and under the base of the pot.

Stand the graft in a warm, shaded spot for a fortnight without disturbing it, and then move it back into full sunlight. It will be a month or so before it is possible to tell whether the graft has taken or not. If it does fail, provided neither graft nor scion show signs of rotting, clean up the two cut surfaces and try again.

One point is important to remember. When grafting a variety which contains no chlorophyll, never bury the green stock, for that is the plant's life-blood!

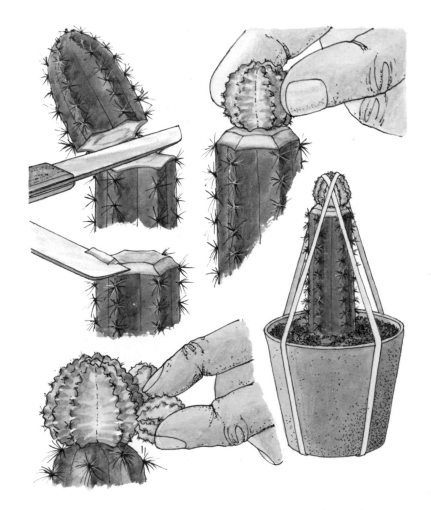

Natural propagation

Many plants have the power to increase themselves by natural means, other than by producing seed. These methods of propagation are:
  Division
  Suckers
  Stolons
  Runners
  Offsets

Division

This is the simplest and most reliable method to increase most hardy herbaceous perennials and many houseplants. It consists of separating a plant into two or more pieces by dividing the crown. (The crown is the part of the plant from which new shoots arise.) In many houseplants the crown comprises tight rosettes of leaves—e.g. saintpaulia (African violet) and sansevieria. The crown of hardy herbaceous perennials can be either a spreading mass of fibrous roots with small shoots—e.g. aster and astilbe—or roots with thick, fleshy buds—e.g. delphinium, paeony and rhubarb. As well as herbaceous perennials and houseplants, a few shrubs—e.g. kerria and philadelphus—and many alpines can be increased quite easily by the method of division.

Outdoor plants are divided in late autumn to early winter, at the end of the growing season, if they are early-flowering. Late summer- and autumn-flowering subjects are best divided in the early spring, just before they start into growth.

Houseplants are divided when they become too large for their pots: this can be at any time of year.

Left: Stages in grafting a cactus. First slice the top of the stock plant, using a sharp knife. Then bevel the edges of the stock plant, taking care not to cut into the ring of vascular tissue. Remove the scion (portion of plant to be grafted on to the stock). Align the scion on the stock, ensuring that their rings of vascular tissue touch in at least one place. Hold the scion securely on the stock by means of two elastic bands.

Above: Increasing saintpaulia (African violet) by division. A large clump is removed from its pot and individual rosettes pulled off. Each rosette must have a good mass of fibrous root attached. Divisions are potted individually into 76-mm (3 in) or 89-mm (3½-in) pots. Other houseplants that can be so increased include sansevieria, aspidistra, most ferns, primula, peperomia, aglaonema and the grass-like *Carex morrowii variegata*.

Above: Clumps of large herbaceous plants can be tough to pull apart. A good way to start is by pushing two digging forks, back to back, right through the clump and then levering it apart. Once in more manageable portions, further dividing can be done by hand or with a small fork. The old woody centre of a large clump should not be replanted—just the young growths which are prised from its perimeter.

Above: Some plants, such as the screw pine (pandanus) and pineapple, throw up numerous shoots around their bases. If these are cut off as near as possible to the parent stem, they can be prepared like cuttings and rooted in cutting compost. If a shoot is produced from below ground level, it will probably have developed roots and should readily grow away into a new plant.

Above: A sucker is a shoot arising from a root, and for propagation purposes is best removed from the parent plant during the dormant season. Dig round the parent plant, exposing the sucker, and cut it off with some roots attached. Trim away any weak shoots and damaged roots, and then either pot or plant the shoot straight into the new site.

Suckers

A sucker is a shoot arising from a bud on a root below ground. Plants that readily sucker include *Rhus typhina* (stag's horn sumach), poplar, lilac, raspberry and rose. Suckers are removed during the dormant season. Dig round the parent plant and cut off the sucker with a sharp knife. This is better than pulling off the suckers, stripping the fibrous roots and damaging the parent plant. Trim any old or weak shoots and pot or plant in the new site. Note that plants which readily sucker can usually also be increased by root cuttings. Be careful not to confuse suckers with long, vigorous, sappy shoots that sometimes develop from the base of trunks; these are 'water-sprouts' and should never be used for propagation.

Stolons

A stolon is a stem that grows prostrate along the ground. It throws down roots, normally from the regions of leaf joints, into the soil. In fact, they can be classed as a type of layer. Stolons can be treated as suckers. When dormant, sever them from the

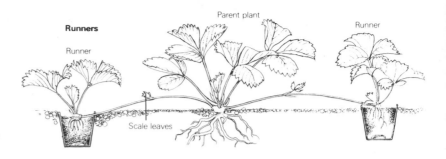

Runners
Runner
Parent plant
Runner
Scale leaves

parent, lift carefully and plant or pot. Plants producing stolons include ajuga, mint, stachys and *Cornus stolonifera,* a rampant spreading dogwood.

Runners

A runner is a specialized long, slender stem, arising from near the base of the parent plant, that produces a new shoot at each leaf joint. It should not be confused with a layer, which is a normal shoot induced to grow roots while attached to the parent plant. Plants producing runners are of the rosette type and include *Saxifraga sarmentosa* (the strawberry geranium) and of course the strawberry. In *Chlorophytum elatum* (spider plant) the flower stalk serves a similar purpose, producing daughter plants after flowering has ceased. Runners are easily rooted by pegging them down into pots filled with cutting compost.

Offsets

An offset is a short, thick side shoot which develops from the

base of a plant—e.g. pineapple. It can also take the form of a thickish stem with a rosette-type shoot at its tip—e.g. sempervium (houseleek), *Saxifraga aizoon* (silver saxifraga) and echeveria. The offshoot, with a small section of stem attached, can be removed and rooted, although it may have formed roots while still connected to the parent plant. Bulbs may also produce offsets and these are described on page 51.

Above: Strawberry plant with runners pegged down, one to a pot, into cutting compost. Prepared in early to mid-summer, runners will be well rooted and ready for planting into permanent positions by autumn for fruiting the following year. Only propagate from healthy, vigorous plants, rooting no more than five runners from each.

Below: Echeveria offsets are borne on thick, fleshy stems covered with scale-like leaves. Rosettes are removed with a short piece of stem attached and quickly root. If an offset has been resting on the ground, it may well develop sufficient roots to be severed and potted as any rooted cutting.

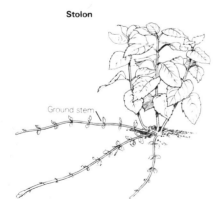

Stolon
Ground stem

Above: Mint, ajuga and stachy are all common plants producing stolons (prostrate stems that grow along the ground and throw down roots). Severed from the parent and potted, a stolon quickly makes a good-sized plant.

Offset
Parent rosette
Scale leaves
Daughter rosette

Monthly work schedules

January

General Buy pots, trays, compost, other propagating equipment, and seed. Wash down the greenhouse, frame, benching and all propagating equipment, including pots and trays. Fumigate greenhouse with a combined insecticide and fungicide to prevent pest and disease problems.

Seed In a propagator, start seed of pelargoniums, fibrous-rooted and tuberous begonias, early tomatoes and gloxinias. In an unheated frame or part of the greenhouse, start bought seeds of trees, shrubs, hardy herbaceous plants and alpines.

Cuttings Take softwood shoot cuttings of carnations and autumn-flowering chrysanthemums, and root cuttings of hardy herbaceous plants.

February

General Send in final orders for seed. Prick out seedlings as soon as they are large enough to handle, and cuttings when rooted. Control rotting of seedlings and cuttings by watering with a copper fungicide or benomyl.

Seed Sow most kinds of bedding plants and other half-hardy and tender plants for outdoor and pot culture (provided your greenhouse can be heated sufficiently at night to prevent losses after germination and pricking out). Aim for a minimum night temperature of 4°C (40°F). Sow greenhouse tomatoes, cucumbers, aubergines and peppers.

Cuttings Take softwood cuttings of bedding plants—e.g. pelargoniums, fuchsias, heliotropes and coleus—and carnations and chrysanthemums. Start dahlia tubers into growth for cuttings.

March

General Fumigate greenhouse once a fortnight from now onwards as a precaution against pests and diseases.

Seed Owners of unheated greenhouses can, with the aid of a propagator, start sowing all types of half-hardy and tender plants. Sow tomatoes, cucumbers, peppers and aubergines for planting in an unheated greenhouse or outdoors. Sow seed of all kinds of houseplants and cacti in a propagator. In a cold frame sow hardy annuals.

Cuttings Continue taking cuttings of bedding plants, carnations and chrysanthemums, and start striking dahlia cuttings. Take softwood cuttings of hardy herbaceous perennials such as lupins and delphiniums. Insert in the open ground hardwood cuttings taken the previous October–December and stored through the winter packed in sharp sand, etc.

Division Increase all types of hardy herbaceous plants by lifting and dividing.

April

Seed Sow melons, marrows, sweet corn and celery, and annuals for use as pot plants in the greenhouse and home during the autumn and winter. Continue with sowings of plants for summer displays outdoors and in the greenhouse. Make sowings of the various primulas for winter displays in the greenhouse and home.

Cuttings Continue taking dahlia cuttings. Take cuttings of all types of houseplants. Start taking softwood cuttings of some shrubs. These will root very quickly with the aid of a mist propagator.

Layering Air layer overgrown houseplants such as crotons, rubber plants and dracaenas. Start simple layering or air layering of trees and shrubs outdoors.

May

Seed Make sowings of greenhouse ornamentals such as asparagus, calceolaria and cyclamen.

Cuttings Take softwood cuttings from trees and shrubs, greenhouse and houseplants, and poinsettias. Take leaf cuttings of sansevieria, *Begonia rex*, saintpaulia (African violet) and peperomia.

Division Divide any large houseplants, such as sansevieria and asparagus, with several stems.

Layering This is the best month to layer trees and shrubs for successful rooting by winter.

June

Seed Sow biennials, such as wallflower, sweet william, and double daisy (bellis) for planting out in spring-flowering quarters in the late autumn. Also try Brompton stocks and forget-me-nots. Sow seed of all hardy herbaceous perennials in the open.

Cuttings Continue taking softwood cuttings of all greenhouse plants, trees and shrubs, including cacti. Increase carnations by 'pippings'.

Division Lift and divide polyanthus, primulas and auriculas and either pot or plant in the open ground.

Grafting Graft cacti.

July

Seed As June.

Cuttings Start taking semi-ripe cuttings of trees and shrubs, and alpines.

Layering Simple layer border carnations. Tip layer blackcurrants.

Bulbs Try increasing hyacinths by basal cuttage.

Runners Peg down strawberry runners to produce new plants.

August

Seed Sow stocks, cyclamen and schizanthus. Outside, sow winter-flowering pansies and violets and stocks for spring flowering.

Cuttings Take pelargonium cuttings from plants growing outdoors: they root easily at this time of year. Continue with shrub cuttings, using semi-ripe wood.

September

Seed Start gathering ripe seed from trees, shrubs and herbaceous plants. Sow hardy annuals outdoors *in situ* for flowering early the following summer, and indoors to give a display in cool greenhouses during the spring.

Cuttings Take cuttings from evergreen trees and shrubs including conifers, and all bedding plants.

October

Seed Sow sweet peas and stand in an unheated frame; otherwise as for September.

Cuttings Start taking hardwood cuttings, either for inserting immediately in an unheated frame or open ground, or for winter storage and inserting in the spring.

Bulbs Increase lilies by scale cuttings.

November

Seed Start sowing begonias and gloxinias, and a late batch of cyclamen. There is still time to start hardy annuals for a late spring or early summer display. Sow seed of trees, shrubs, etc., and place containers in a cold frame.

Cuttings Continue to take hardwood cuttings. Rose cuttings can be prepared from pruning wood. Prepare and insert root cuttings of hardy herbaceous plants, trees and shrubs.

December

General Send away for seed catalogues. Start fumigating the greenhouse regularly to keep pest and disease problems in check.

Seed As for November.

Cuttings Start taking cuttings from carnations and chrysanthemums. In an unheated frame insert cuttings of border carnations.

Pest and disease control

Seeds, seedlings, cuttings, divisions, grafts and layers are all susceptible to damage by a host of diseases and pests, as the table opposite clearly illustrates.

While most pests and diseases can be controlled, it is far more satisfactory if they can be prevented from occurring in the first place. A system of total sanitation, as achieved by many commercial growers, is hardly practical when working with a small greenhouse. However, simple precautions can be taken.

First and foremost, ensure that all materials—tools, pots, wire, trays, propagators, watering equipment, etc.—are cleaned before use. Household bleach or disinfectant is suitable, or formalin can be used at the rate of one part formalin (40 per cent formaldehyde) to 50 parts water. It should be remembered that formalin can burn the skin, so gloves should be worn, and it must not come into contact with the eyes.

Garden debris, including old potting compost, dead or dying plants, etc., is another source of infection and all such refuse should be cleared out. Wash benching down with a disinfectant at least once a year and keep glass surfaces clean. Never allow weeds to grow inside a greenhouse; many are host plants for pests and diseases. Similarly, keep the immediate surrounds of the greenhouse clear of rubbish and weeds. Plant debris should be placed on a compost pile away from the greenhouse and all diseased material should be burnt. Always use sterilized growing mediums.

As a precautionary measure, fumigate the greenhouse regularly with an insecticide and fungicide such as tecnazene and HCH (also known as lindane and BHC). Both can be bought as smokes (so that all you need to do is light the touch-paper and retire) or as volatile preparations for use in a special electric vaporizer.

When selecting material for propagation, ensure that it is free from damage. Tissue should not be soft, bloated, weak or spindly. Cut surfaces should, wherever practical, be protected with a fungicide dust to prevent disease from entering. Rooting cuttings in pots means that outbreaks of disease can be confined. Pick over cuttings weekly, removing dead and dying leaves and cuttings.

With seed-sowing and seedlings the points to watch are:
1 Overcrowding caused by sowing and pricking out too thickly
2 Watering with contaminated water from a butt
3 Over-watering
4 Too low temperatures and very wet soil conditions during winter months
5 Poor light immediately after germination
6 Very humid growing conditions after germination.

Left: Potted seedlings showing symptoms of damping-off, a shrinking and rotting of seedling stems at soil level. It results from attack by certain soil-borne fungi.

Above: Plant suffering from blackleg, a disease caused by several fungi. This results in the stems of pelargoniums, and some other cuttings, decaying and turning black below soil level.

Left: As a precaution against pests and diseases fumigate your greenhouse regularly with an insecticide and a fungicide.

Problem and cause	Plants attacked	Description and damage	Prevention and control
Blackleg —various soil bacterial and fungal organisms	Most types of cuttings, especially pelargonium and begonia	Blackening of base of cutting followed by wilting of foliage and death	Use sterilized compost for rooting cuttings. Treat cuttings with fungicide before inserting. Avoid taking cuttings in late autumn or winter. Treat cuttings while rooting with a copper fungicide. Destroy any cuttings showing slightest sign of infection
Club root —fungal organism	All members of crucifer family— e.g. cabbage, cauliflower, and wallflower	Only attacks plants being raised in the open ground. Roots swell, distort and fail to grow properly, resulting in failure of top growth	Do not grow plants in contaminated soil unless it is treated with calomel. Burn roots of all plants likely to be infected. A problem in the greenhouse only when contaminated soil is used in seed and potting composts
Cutworms —larvae of various insects	Young plants and seedlings	Eat plants at soil level, causing them to wilt or topple over	Use sterilized compost in the greenhouse. Dust soil outdoors with HCH or a proprietary soil insecticide
Damping off —several soil fungal organisms	All seedlings raised in containers	Stem tissue invaded at, or just below, soil level. Seedlings topple over and die	Sow seed thinly in sterilized compost. Treat seed with thiram prior to sowing. Maintain good ventilation, avoid over-watering. Water compost with Cheshunt compound. If disease is present, water with Cheshunt compound or benomyl, or dust with thiram, captan or sulphur and remove infected seedlings
Grey mould —fungus	Plants in propagation cases, etc.	Grey mould develops on surface of infected plants. Attacks usually start on damaged or dead plant material. Thrives in moist, humid conditions	Regularly ventilate propagating cases. Pick off dead and dying leaves. Remove infected cuttings and destroy. Spray plant material with benomyl or captan, or dust with captan or sulphur. Avoid overcrowding
Mice	Bulbs, etc., and large seeds	Dig up and eat material	Use traps or put down poisonous bait
Millipedes —insect-like pests with many legs	Seedlings and small bulbs, etc.	Gnaw any soft plant tissue, causing small cavities in bulbs, and seedlings to topple over	Dust compost or soil with HCH or a proprietary soil insecticide. Use sterilized compost
Seed rot —many fungal and bacterial organisms	Seeds	Seeds sown in cold, damp soil are most prone, especially winter sowings outdoors	Avoid over-wet composts. Use sterilized compost. Dust seed with thiram or captan prior to sowing
Slugs and snails	All plants	Leaves and stems eaten, often to soil level. Slime trails nearly always present	Use proprietary slug bait or water with metaldehyde
Springtails —minute wingless, jumping insects	Bulbs, etc., and seedlings	Feed on dead material, seedling stems and surface of bulbs, etc. Seedling leaves often pitted and small roots sometimes eaten	Encouraged by wet conditions, so avoid over-watering. Dust compost and plants with HCH or water with malathion
Symphylids —small white elongated insects	All young plants and seedlings	Attack at soil level, feeding on all types of living material. Seedlings frequently killed	Encouraged by warm, moist conditions. Tend to collect in propagators, burying themselves in the compost. Apply drenches of HCH
Woodlice —crustaceans	All soft plant material	Tend to live in rotting material and feed at night. Roots frequently gnawed	Dust with carbaryl or HCH. Remove all rubbish likely to provide hiding places

Glossary

Adventitious roots: Roots normally developing from part of plant other than primary roots—e.g. stem or leaf.

Asexual propagation: Increasing plants by vegetative means—e.g. stem and root cuttings.

Back bulb: Old, leafless pseudo-bulb (see page 54) that can be used to propagate some orchids.

Blackleg: Black rotting of base of cutting caused by a variety of soil-borne fungal and bacterial organisms.

Bleeding: Loss of sap from the cut base of a cutting or wound in a plant.

Bottom heat: Heat supplied to propagation material, such as seeds and cuttings, by warming the rooting medium.

Broadcasting: Sowing seeds by scattering them over a wide strip of ground in a random manner.

Bud: Compact, undeveloped shoot consisting of a much shortened stem with leaves and/or embryo flower, and usually protected by scale-like leaves.

Budding: Form of grafting (see page 55) whereby a bud cut from a shoot is grafted on to a root-stock.

Bulb: Modified shoot consisting of a very shortened, almost flat, stem enclosed by fleshy, scale-like leaves.

Bulbil: Small bulb produced in the axil of a leaf above ground.

Bulblet: Small bulb that develops on a stem at or below ground level.

Bulb scale: Fleshy, modified leaf of a bulb sometimes used for propagation (see page 52).

Callus: 'Scab' tissue, often white, developing on or around a wound.

Chitting: Treatment applied by hand to seed prior to sowing, to encourage germination.

Clove: One offset (see page 51) of a garlic bulb.

Compost: Specially-prepared mixture composed of various ingredients. Used for sowing and germination of seeds, rooting of cuttings and growing plants in containers.

Corm: Short swollen stem containing food and bearing buds in the axils of scale-like leaves which form a protective cover.

Cormel: Miniature corm developing around the base of a corm.

Cormlet: See Cormel.

Cotyledon: Leaf or leaves forming part of the embryo plant. May be carried above ground after germination to act as first foliage or remain below the surface. Normally serves as a food store.

Crown: Part of the plant at ground level where stems meet roots and new shoots normally arise from buds.

Cutting: Portion of plant—stem, root, leaf or shoot—with the ability to regenerate all the organs needed to form a complete new plant.

Damping-off: Soil-borne fungal disorder that causes rotting, collapse and death of seedlings by attacking stems at or below ground level.

Disbudding: The removal of unwanted buds.

Division: Separating a plant into two or more portions by dividing the crown.

Dormancy: State where plant or seed is alive but not growing. Usually brought about by unfavourable growing conditions such as low temperature or lack of moisture.

Drill: Shallow channel or groove made in the soil for receiving seed.

Embryo: Young plant contained within the seed.

Etiolation: Condition where stems become abnormally long and weak and leaves small and pale from lack of light.

Eye: 1. Cutting consisting of a small piece of stem containing a single bud.
2. Bud on a stem tuber such as a potato.

F_1 (first filial generation): First generation of offspring resulting from crossing two distinct parent plants.

Fertilization: Fusion of male and female sex cells, resulting in a single cell which develops into the seed.

Germination: Emergence of the embryo plant from the seed and its growth and development into a seedling.

Graft: Union of two portions of plants to form a single plant.

Green bulb: Pseudobulb with leaves.

Hardening off: Gradual acclimatization of plants from the tender environment of a greenhouse to harsher outdoor conditions.

Heel cutting: Cutting taken with a slither of older tissue attached to its base.

Hormone rooting compound: Synthetic growth regulator normally applied to the base of a cutting to encourage more rapid root development.

Hybrid: Plant resulting from crossing two parents that are genetically different.

Internode: Portion of stem between two leaf joints or nodes.

John Innes composts: Standardized range of soil-based composts.

Layer: A stem induced to develop roots while it is attached to the parent plant.

Light: Glazed sash used to cover a frame.

Mist propagation: Method of propagation, usually for cuttings, where plant material is kept constantly damp by misting with water.

Node: Part of a stem where one or more leaves arise.

Offsets: New bulbs which develop and separate from the mother bulb.

Pelleted seed: Small seed encased in a layer of clay-like material to enable precise sowing and reduce waste.

Pipping: Type of cutting prepared by pulling shoot tip out of a node.

Plant out: Transfer of plant to a new growing position, which is normally outdoors in the open ground.

Pot-off, pot-up: Transfer of seedling, newly-rooted cutting or young plant from propagating bench or seed tray to a pot.

Pot-on: Transfer of a pot-grown plant into a larger container.

Prick-out: Transfer of seedlings from their place of germination to pots, boxes or the open ground to allow them more growing space.

Prothallus: Minute heart-shaped structure bearing a fern's sexual organs.

Pseudobulb: Swollen sections of stem produced at or above ground level by some orchids.

Rhizome: Specialized stem with scale-like leaves growing at or below ground level.

Rootstock: Bottom part of a grafted plant which forms the root system.

Runner: Specialized long, slender stem, arising from near the base of the parent plant, that produces a new shoot at each leaf joint.

Scarification: Deliberate damage of seed coat by hand to encourage germination. See also Chitting.

Scion: Part of a graft which develops into the new plant.

Scooping: Removal of the entire basal plate from a bulb to expose the base of the storage leaves.

Scoring: Making deep cuts through the basal plate of a bulb.

Seed: Embryo plant packed within a hard protective coat to keep it from being damaged until it germinates.

Seed leaf: See Cotyledon.

Seedling: Young plant raised from a seed.

Sexual reproduction: Reproduction by fusion of male and female sex cells to give rise to a seed.

Soil block: Small cube of compressed compost which acts as a complete growing medium and pot for a single plant.

Soil warming: See Bottom heat.

Space sowing: Sowing seed at constant distances in either drills or containers.

Spore: Microscopic seed-like structure capable of giving rise to a new plant. See page 25.

Stratification: Exposing seed to moist, cold (but not freezing) conditions in order to induce germination.

Stock plant: Selected plant grown as a source of material for cuttings.

Stolon: Stem that grows prostrate along the ground, putting down roots, normally from the regions of leaf joints, into the soil.

Strike: To root (a cutting).

Sucker: Shoot arising from a bud on a root below ground.

Tap root: Root system with a thick, prominent main root.

Thinning: Reducing the number of plants (in a row for example), leaving the remainder more room to develop.

Transplanting: The act of transferring a plant from one position to another.

Tuber: Swollen end of an underground stem bearing buds (e.g. potato) or swollen root with buds only on the crown (e.g. dahlia).

Tubercle: Small aerial tuber borne in the axil of a leaf (e.g. *Begonia sutherlandii*).

Vegetative propagation: See Asexual propagation.

Water shoot or 'sprout': Vigorous, sappy shoot growing from a branch or trunk of a tree.

Weaning: Controlled transfer of plants from a propagator to an open greenhouse bench without inducing a serious shock to growth.

Wilt: Loss of turgidity in a plant from lack of water.

Wounding: Injuring the base of a cutting, usually by removing a slither of rind, to induce root development.

Index

Acknowledgements

Artists
David Briant/Joan Farmer Artists
Ron Hayward Associates

Photographs
John Harris 8, 19, 21, 22, 36, 43, 60
Humex Ltd 7, 9, 12, 13, 16
Michael Warren 1, 5, 8, 20, 27, 28, 29, 32, 33, 37, 38, 42, 44, 55